Cyanobacteria - Exploring Their Role in Energy, Environment and Industry

Edited by Ihana Aguiar Severo and Mariane Bittencourt Fagundes

Published in London, United Kingdom

Cyanobacteria - Exploring Their Role in Energy, Environment and Industry
http://dx.doi.org/10.5772/intechopen.1008609
Edited by Ihana Aguiar Severo and Mariane Bittencourt Fagundes

Contributors
Bharat Kumar Majhi, Claudia Hoepfner, Colin Gates, Daniel Guzmán, Flavio Oliveira, Graciliana Lopes, Guilherme Scotta Hentschke, Gustavo Chavez, Hamse Kjerstadius, Hana Faulds, Hugo Pereira, Jing Li, Kritika Tripathi, Leslie Castillo, Maria do Rosário Domingues, Mariane Bittencourt Fagundes, Marisa Pinho, Moshe Gophen, Nicole Seliga, Nidhi Patel, Sara Ferreira, Tiago Conde, Tánia Melo, Vitor Vasconcelos, Yogesh Mishra, Íhana Aguiar Severo

First published in London, United Kingdom, 2026 by IntechOpen
IntechOpen is the global imprint of INTECHOPEN LIMITED, registered in England and Wales, registration number: 11086078, 167-169 Great Portland Street, London, W1W 5PF, United Kingdom

For EU product safety concerns: IN TECH d.o.o., Prolaz Marije Krucifikse Kozulić 3, 51000 Rijeka, Croatia, info@intechopen.com or visit our website at intechopen.com.

British Library Cataloguing-in-Publication Data
A catalogue record for this book is available from the British Library

Cyanobacteria - Exploring Their Role in Energy, Environment and Industry
Edited by Ihana Aguiar Severo and Mariane Bittencourt Fagundes
p. cm.
Print ISBN 978-0-85014-924-1
Online ISBN 978-0-85014-923-4
eBook (PDF) ISBN 978-0-85014-925-8

If disposing of this product, please recycle the paper responsibly.

Meet the editors

Ihana Aguiar Severo is a researcher at the Federal University of Paraná (Sustainable Energy Research and Development Center, NPDEAS) and at Florida State University (Center for Advanced Power Systems, CAPS). She holds a Ph.D. (2020) and an MSc (2017) in Food Science and Technology from the Federal University of Santa Maria, Brazil. She is the author of more than 100 scientific contributions, including peer-reviewed journal articles, book chapters, edited and authored books, and international conference papers. Her research focuses on microalgae biotechnology, photobioreactors, environmental applications, and renewable energy systems.

Mariane Bittencourt Fagundes is a researcher specializing in microalgae and cyanobacteria biotechnology. She works at the Interdisciplinary Centre for Marine and Environmental Research (CIIMAR) at the University of Porto, Portugal. She is currently part of the BlueBio4Future team researching bioactive metabolites produced by cyanobacteria and their biotechnological applications. She holds a Ph.D. (2020) and an MSc (2017) in Food Science and Technology from the Federal University of Santa Maria, Brazil, including a research stay at the Institute of Food Science Research (CIAL) of the Autonomous University of Madrid. Her research focuses on the production, extraction, and characterization of bioactive compounds from cyanobacteria, especially terpenes, sterols, and other secondary metabolites, using advanced metabolomics approaches, gas chromatography, and mass spectrometry.

Contents

Preface

Cyanobacteria are among the most ancient and influential organisms on Earth, playing a fundamental role in shaping the planet's atmosphere, ecosystems, and biological evolution. Today, these microorganisms are gaining renewed attention for their exceptional potential to address contemporary challenges in energy production, environmental sustainability, and industrial biotechnology.

This edited volume, *Cyanobacteria – Exploring Their Role in Energy, Environment and Industry,* brings together contributions from experts across disciplines to provide a comprehensive overview of cyanobacterial research, ranging from fundamental biological processes to selected applied perspectives. The book covers key topics, including i) an introduction to cyanobacteria, ii) biological aspects of cyanobacteria for industrial applications, iii) structural and functional roles of inorganic cofactors of photosystem II, iv) genetic modification strategies, v) algae-based biorefineries, vi) commercial potential of cyanobacteria through biochemical and taxonomical analysis, and vii) ecological dynamics influencing cyanobacterial dominance in natural systems.

This volume emphasizes foundational concepts, biological mechanisms, and representative case studies that illustrate the diverse roles of cyanobacteria in natural and engineered systems. The chapters provide a solid scientific basis for understanding cyanobacterial physiology, genetics, ecological dynamics, and emerging applications, serving as a reference for readers seeking a comprehensive overview of the field. This book is intended for researchers, graduate students, and professionals interested in cyanobacteria from biological, environmental, and applied science perspectives.

Dr. Ihana Aguiar Severo
Federal University of Paraná,
Sustainable Energy Research and Development Center (NPDEAS),
Curitiba, PR, Brazil

Dr. Mariane Bittencourt Fagundes
CIIMAR - Interdisciplinary Centre of Marine and
Environmental Research of the University of Porto,
Matosinhos, Portugal

Chapter 1

Introductory Chapter: Cyanobacteria – Exploring Their Role in Energy, Environment and Industry

Íhana Aguiar Severo and Mariane Bittencourt Fagundes

1. Overview of cyanobacteria: ancient microorganisms with modern relevance

Cyanobacteria are among the most ancient life forms on Earth, with fossil records dating back more than 3.5 billion years. Through oxygenic photosynthesis, these microorganisms fundamentally transformed the planet's atmosphere, paving the way for the evolution of aerobic life and shaping global biogeochemical cycles. Despite their prokaryotic cellular organization, cyanobacteria exhibit remarkable metabolic diversity and ecological adaptability, thriving in freshwater, marine, and terrestrial ecosystems worldwide. Their evolutionary success and versatility have attracted increasing attention for biotechnological applications focused on sustainability and resource efficiency [1, 2].

Unlike eukaryotic microalgae, cyanobacteria possess a prokaryotic cell structure and, in many species, the ability to fix atmospheric nitrogen [3]. These characteristics make them particularly attractive as biological factories capable of generating a wide range of bio-based products using sunlight, water, and CO_2 as the main resources. To date, more than 2,000 bioactive metabolites have been identified from cyanobacterial species, underscoring their biochemical richness and industrial relevance [4].

From an evolutionary perspective, this versatility is strongly linked to the cyanobacterial pangenome, which encompasses the extensive and diverse gene pool distributed across different lineages. Rather than relying solely on gradual mutations, cyanobacteria rapidly adapt to environmental changes through genome fluidity and horizontal gene transfer, particularly within heterogeneous genomic regions [5]. Mechanisms such as transduction, conjugation, and transformation have operated throughout their evolutionary history, enabling gene exchange across lineages for billions of years [6].

As a result, cyanobacterial genomes often display a mosaic architecture, where genetic innovation arises from both inherited traits and acquired functions. This evolutionary resilience has allowed cyanobacteria to survive major planetary transitions, including the Great Oxygenation Event and multiple mass extinction episodes, while continuously expanding into new ecological niches [7].

2. Cyanobacteria in the context of sustainability and the bioeconomy

As the global community transitions toward a low-carbon future, cyanobacteria have emerged as promising contributors to the bioeconomy. Their ability to produce bioenergy and bioproducts without competing with food crops offers a sustainable path for circular resource utilization. Cyanobacterial systems can be integrated into biorefineries that couple waste valorization with biomass generation, contributing to carbon capture and nutrient recycling. The cultivation of microalgae and cyanobacteria using wastewater and digestate is central to developing circular biorefineries, as emphasized by recent reviews [8–12].

The metabolic flexibility of cyanobacteria enables the synthesis of biofuels such as biohydrogen, biodiesel precursors, and biogas [13, 14]. Moreover, they produce valuable compounds like pigments (phycocyanin, chlorophyll, carotenoids) [15–17], bioplastics (polyhydroxyalkanoates) [18], biofertilizers [19], pharmaceuticals [20], and biomedical applications [21]. In addition to well-known pigments and polymers, cyanobacteria synthesize lipid-derived metabolites such as sterols and squalene, which play important roles in membrane structure, cellular adaptation, and stress tolerance. These compounds have been identified and produced in strains such as *Phormidium autumnale*, highlighting their potential as high-value products for nutraceutical, pharmaceutical, and biorefinery applications within sustainable cyanobacteria-based systems [22–24]. All of these applications align with circular economy principles, converting waste and emissions into renewable resources while reducing environmental impacts [25].

3. Environmental implications and biotechnological challenges

Cyanobacteria play a dual role in environmental systems: they are vital for ecosystem balance and carbon cycling, but they can also cause ecological concerns such as harmful algal blooms and toxin production [26–28]. Understanding and managing these dualities is central to sustainable biotechnological development. Advances in molecular biology, genomics, and environmental engineering now enable precise control over cyanobacterial metabolism and growth conditions [29–31].

Cultivation technologies have evolved from open ponds to closed photobioreactors, improving process control and product quality. However, challenges remain in large-scale cultivation, including contamination risks, harvesting efficiency, and the optimization of light and nutrient supply [32]. Future research aims to enhance strain robustness and productivity through synthetic biology and metabolic engineering.

4. Industrial and technological advances

Recent developments in biotechnology have expanded cyanobacteria's industrial applications. They are being engineered to function as biofactories for pigments, antioxidants, amino acids, and bioactive metabolites [33–37]. Their

DOI: http://dx.doi.org/10.5772/intechopen.1014612

capacity to fix nitrogen naturally also makes them suitable for sustainable agriculture and biofertilizer production. In addition, their use as feedstock for bio-based materials and biodegradable polymers positions them as a strategic platform for replacing petroleum-based products [38].

Integration with industrial systems is progressing through biorefineries and wastewater treatment plants, where cyanobacteria provide multiple benefits: bioremediation, biomass recovery, and production of high-value compounds [38]. The potential for coupling cyanobacterial processes with renewable energy systems, such as solar-driven photobioreactors, further strengthens their industrial appeal.

5. Future perspectives

The next frontier for cyanobacterial biotechnology lies in digitalization, automation, and predictive modeling of bioprocesses. Combining artificial intelligence with omics-based data can accelerate strain selection, optimize

Figure 1.
Multifunctional roles of cyanobacteria across evolutionary, environmental, and industrial dimensions (prepared using an AI tool).

metabolic fluxes, and enable continuous monitoring of large-scale systems. Furthermore, hybrid systems integrating cyanobacteria with other microorganisms are emerging as powerful platforms for multi-product biomanufacturing.

In the context of global sustainability, cyanobacteria contribute directly to several United Nations Sustainable Development Goals (SDGs), including clean energy (SDG 7), climate action (SDG 13), and responsible production and consumption (SDG 12). Their versatility, scalability, and environmental benefits make them key biological agents in building a green industrial future. **Figure 1** provides an overview of the roles of cyanobacteria across evolutionary, environmental, and industrial dimensions.

Acknowledgment

The author acknowledges the use of ChatGPT 5.2 for language polishing of the chapter.

Author details

Íhana Aguiar Severo[1*] and Mariane Bittencourt Fagundes[2]

1 Sustainable Energy Research & Development Center (NPDEAS), Federal University of Paraná (UFPR), Curitiba, Brazil

2 CIIMAR- Interdisciplinary Center of Marine and Environmental Research, University of Porto, Porto, Portugal

*Address all correspondence to: ihana.aguiar@gmail.com

References

[1] Severo IA, Martínez-Burgos WJ, Ordonez JC. Insights into Algae-Fundamentals, Culture Techniques and Biotechnological Uses of Microalgae and Cyanobacteria. London: Intechopen; 2024. DOI: 10.5772/intechopen.1001726

[2] Cameron ES, Sanchez S, Goldman N, et al. Diversity and specificity of molecular functions in cyanobacterial symbionts. Scientific Reports. 2024;**14**:18658

[3] Satta A, Esquirol L, Ebert B. Current metabolic engineering strategies for photosynthetic bioproduction in cyanobacteria. Microorganisms. 2023;11. DOI: 10.3390/microorganisms11020455

[4] Jones MR, et al CyanoMetDB, a comprehensive public database of secondary metabolites from cyanobacteria. Water Research. 2021;**196**:117017

[5] Tokuda M, Shintani M. Microbial evolution through horizontal gene transfer by mobile genetic elements. Microbial Biotechnology. 2024;17. DOI: 10.1111/1751-7915.14408

[6] Desai M, Walczak A. Flexible gene pools. Science. 2015;**348**:977–978. DOI: 10.1126/science.aab3957

[7] Boc A, Makarenkov V. Towards an accurate identification of mosaic genes and partial horizontal gene transfers. Nucleic Acids Research. 2011;**39**: e144–e144. DOI: 10.1093/nar/gkr735

[8] Cheirsilp B, Maneechote W, Srinuanpan S, Angelidaki I. Microalgae as tools for bio-circular-green economy: Zero-waste approaches for sustainable production and biorefineries of microalgal biomass. Bioresource Technology. 2023;**387**:129620. DOI: 10.1016/j.biortech.2023.129620

[9] Ezhumalai G, Arun M, Manavalan A, Rajkumar R, Heese K. A holistic approach to circular bioeconomy through the sustainable utilization of microalgal biomass for biofuel and other value-added products. Microbial Ecology. 2024;87. DOI: 10.1007/s00248-024-02376-1

[10] Gurreri L, Rindina M, Luciano A, Lima S, Scargiali F, Fino D, Mancini G. Environmental sustainability of microalgae-based production systems: Roadmap and challenges towards the industrial implementation. Sustainable Chemistry and Pharmacy. 2023;**35**:101191. DOI: 10.1016/j.scp.2023.101191

[11] Tsapekos P, Khoshnevisan B, Alvarado-Morales M, Zhu X, Pan J, Tian H, Angelidaki I. Upcycling the anaerobic digestion streams in a bioeconomy approach: A review. Renewable and Sustainable Energy Reviews. 2021;**151**:111635. DOI: 10.1016/j.rser.2021.111635

[12] Wicker R, Kumar G, Khan E, Bhatnagar A. Emergent green technologies for cost-effective valorization of microalgal biomass to renewable fuel products under a biorefinery scheme. Chemical Engineering Journal. 2021;**415**:128932. DOI: 10.1016/j.cej.2021.128932

[13] Jacob-Lopes E, Santos AB, Severo IA, Deprá MC, Maroneze MM, Zepka LQ. Dual production of

bioenergy in heterotrophic cultures of cyanobacteria: Process performance, carbon balance, biofuel quality and sustainability metrics. Biomass and Bioenergy. 2020;**142**:105756. DOI: 10.1016/j.biombioe.2020.105756

[14] Singh V, Jha S, Rana P, Soni R, Lalnunpuii R, Singh P, Sinha R, Singh G. Cyanobacteria as a biocatalyst for sustainable production of biofuels and chemicals. Energies. 2024;**17**:408. DOI: 10.3390/en17020408

[15] Aoki J, Koshikawa R, Asayama M. Recent progress in the cyanobacterial products and applications of phycocyanins. World Journal of Microbiology and Biotechnology. 2025;41. DOI: 10.1007/s11274-025-04297-8

[16] Senatore V, Rueda E, Català M, Díez-Montero R, Ferrer I, Zarra T, Naddeo V, García J. Production of phycobiliproteins, bioplastics and lipids by the cyanobacteria Synechocystis sp. treating secondary effluent in a biorefinery approach. The Science of the Total Environment. 2022;159343. DOI: 10.1016/j.scitotenv.2022.159343

[17] Zittelli G, Lauceri R, Faraloni C, Benavides A, Torzillo G. Valuable pigments from microalgae: Phycobiliproteins, primary carotenoids, and fucoxanthin. Photochemical & Photobiological Sciences. 2023;1–57. DOI: 10.1007/s43630-023-00407-3

[18] Sun L, García J, Gonzalez-Flo E. Potential sustained production of polyhydroxyalkanoate copolymers using cyanobacteria microbiomes. Scientific Reports. 2025;15. DOI: 10.1038/s41598-025-12254-6

[19] Pathak JR, Maurya P, Singh S, Häder D, Sinha R. Cyanobacterial farming for environment friendly sustainable agriculture practices: Innovations and perspectives. Frontiers in Environmental Science. 2018;6. DOI: 10.3389/fenvs.2018.00007

[20] Quintana N, Van Der Kooy F, Van De Rhee M, Voshol G, Verpoorte R. Renewable energy from Cyanobacteria: Energy production optimization by metabolic pathway engineering. Applied Microbiology and Biotechnology. 2011;**91**:471–490. DOI: 10.1007/s00253-011-3394-0

[21] Severo IA, Dias RR, Do Nascimento TC, et al Microalgae-derived polysaccharides: Potential building blocks for biomedical applications. World Journal of Microbiology and Biotechnology. 2022;**38**:150

[22] Fagundes, et al Insights in cyanobacteria lipidomics: A sterols characterization from Phormidium autumnale biomass in heterotrophic cultivation. Food Research International. 2019;**119**:777–784

[23] Fagundes MB, Vendruscolo RG, Maroneze MM, et al Towards a sustainable route for the production of squalene using cyanobacteria. Waste and Biomass Valorization. 2019;**10**:1295–1302

[24] Fagundes, et al. Phytosterol-rich compressed fluids extracts from Phormidium autumnale cyanobacteria with neuroprotective potential. Algal Research. 2021;**102264**. DOI: 10.1016/j.algal.2021.102264

[25] Gradíssimo D, Xavier L, Santos A. Cyanobacterial polyhydroxyalkanoates: A sustainable alternative in circular economy. Molecules. 2020;**25**. DOI: 10.3390/molecules25184331

[26] Joseph J, Ray J. A critical analysis of ecology and diversity of Cyanobacteria in long-term chemicalized tropical plantation field soils: Implications for sustainable soil fertility management. Journal of Environmental Management. 2025;**392**:126817. DOI: 10.1016/j.jenvman.2025.126817

[27] Lucius S, Hagemann M. The primary carbon metabolism in cyanobacteria and its regulation. Frontiers in Plant Science. 2024;15. DOI: 10.3389/fpls.2024.1417680

[28] Victoria A, Astbury M, McCormick A. Engineering highly productive cyanobacteria towards carbon negative emissions technologies. Current Opinion in Biotechnology. 2024;**87**:103141. DOI: 10.1016/j.copbio.2024.103141

[29] Huisman J, Codd G, Paerl H, Ibelings B, Verspagen J, Visser P. Cyanobacterial blooms. Nature Reviews Microbiology. 2018;**16**:471–483. DOI: 10.1038/s41579-018-0040-1

[30] Rastogi R, Madamwar D, Incharoensakdi A. Bloom dynamics of cyanobacteria and their toxins: Environmental health impacts and mitigation strategies. Frontiers in Microbiology. 2015;6. DOI: 10.3389/fmicb.2015.01254

[31] Zhang C, Zhou C, Burnap R, Peng L. Carbon/nitrogen metabolic balance: Lessons from cyanobacteria. Trends in Plant Science. 2018;**23**(12):1116–1130. DOI: 10.1016/j.tplants.2018.09.008

[32] Novoveská L, Nielsen SL, Eroldoğan OT, Haznedaroglu BZ, Rinkevich B, Fazi S, Robbens J, Vasquez M, Einarsson H. Overview and challenges of large-scale cultivation of photosynthetic microalgae and cyanobacteria. Marine Drugs. 2023;**21**(8):445. DOI: 10.3390/md21080445

[33] Ng I, Keskin B, Tan S. A critical review of genome editing and synthetic biology applications in metabolic engineering of microalgae and cyanobacteria. Biotechnology Journal. 2020;**15**:e1900228. DOI: 10.1002/biot.201900228

[34] Pagels F, Vasconcelos V, Guedes A. Carotenoids from cyanobacteria: Biotechnological potential and optimization strategies. Biomolecules. 2021;11. DOI: 10.3390/biom11050735

[35] Singh R, Parihar P, Singh M, Bajguz A, Kumar J, Singh S, Singh V, Prasad S. Uncovering potential applications of cyanobacteria and algal metabolites in biology, agriculture and medicine: Current status and future prospects. Frontiers in Microbiology. 2017;8. DOI: 10.3389/fmicb.2017.00515

[36] Verma S, Thapa S, Siddiqui N, Chakdar H. Cyanobacterial secondary metabolites towards improved commercial significance through multiomics approaches. World Journal of Microbiology and Biotechnology. 2022;38. DOI: 10.1007/s11274-022-03285-6

[37] Zahra Z, Choo D, Lee H, Parveen A. Cyanobacteria: Review of current potentials and applications. Environments. 2020;**7**:13. DOI: 10.3390/environments7020013

[38] Singh J, Kumar A, Rai A, Singh D. Cyanobacteria: A precious bio-resource in agriculture, ecosystem, and environmental sustainability. Frontiers in Microbiology. 2016;7. DOI: 10.3389/fmicb.2016.00529

Chapter 2

Cyanobacteria: A Biological Chassis for Industrial Applications

Bharat Kumar Majhi

Abstract

Cyanobacteria are a diverse group of prokaryotic organisms that engage in oxygenic photosynthesis. They have served as model organisms for a wide range of photobiochemical research. Recently, they have received significant attention as a potential host for various industrial and agricultural applications due to their unique physio-biochemical properties. Their ability to withstand harsh environmental conditions and internal biochemical changes, along with their faster growth, minimal growth requirements, inherent double homologous recombination, and uptake of exogenous deoxyribonucleic acid (DNA) without the use of external forces, makes them a superior candidate for numerous applications. They have been utilized as hosts to manufacture a wide range of industrial products, including biofuel and bioplastic. Furthermore, cyanobacteria's nitrogen-fixing ability enables them to be used as a biofertilizer in sustainable agriculture. Additionally, their ability to scavenge metals from the environment makes them an excellent candidate for bioremediation. This book chapter will explore cyanobacterial applications in biofuel and bioplastic production, as well as bioremediation and sustainable agriculture. It will also highlight their distinct physiological characteristics that differentiate them from other photoautotrophs.

Keywords: cyanobacteria, biofuel, bioplastic, bioremediation, biofertilizer

1. Introduction

Advanced recombinant-DNA technology and synthetic biology have been effective tools for creating genetically modified organisms capable of producing a wide variety of industrial products [1, 2]. Numerous native organisms, including mammals, yeast, *Escherichia coli*, plants, algae, and cyanobacteria, have been genetically altered and transformed into biofactories capable of producing valuable chemicals [3]. Genetically modified host organisms are critical to the current bio-based product generation process. Although various organisms are capable of producing valuable industrial products, growth rate, biomass yield, contamination risks, environmental constraints, production cost, and yield, all play important roles in selecting the right host for biomanufacturing. Mammalian cells are heavily used in a variety of applications; however, contamination risks and processing costs pose significant challenges to bioproduct generation [4]. In contrast, fermentative organisms such as yeast and *E. coli* offer numerous benefits as hosts, including faster growth, higher product yields, and greater resistance to contamination [5]. Despite their many advantages, using fermentative organisms as hosts to produce biochemicals incurs high production costs. Their growth is dependent on external energy sources, including expensive growth media. Furthermore, they are prone to contamination due to their nutrient-rich growth media [5].

To alleviate the aforementioned challenges, photosynthetic organisms have recently been considered as potential hosts for biomanufacturing [6–8]. In addition to being hosts for bioproduct generation, they have also been used in a variety of applications, such as bioremediation and sustainable agriculture, due to their unique physiological properties [9]. They do not require expensive growth media for cultivation, as they are photoautotrophs and synthesize organic carbon from atmospheric carbon dioxide (CO_2) using sunlight and water, which they use for their growth. Moreover, they are resistant to a wide range of pathogens, including zoonotic contaminants. However, photosynthetic organisms, such as plants and algae, have delayed physiological growth, a complex genetic reconstruction process, and low yield, making them less desirable as hosts for bioproduct generation and other applications [10, 11]. Cyanobacteria, on the other hand, have numerous advantages as hosts compared to other host species [12].

Cyanobacteria are ancient photoautotrophic organisms that have the ability to convert CO_2 to organic carbon through photosynthesis [13]. They are single-celled prokaryotic organisms that are better suited for genetic modifications due to their inherent ability to uptake exogenous DNA into the cell and integrate it into the genome through direct homologous recombination [14]. In addition, they grow faster compared to other photoautotrophs and have minimal growth requirements. They can be grown in wastewater, which can reduce cultivation costs [15]. Unlike plants, they do not require arable land to grow. They can be grown outside in low-cost open ponds or indoors in a photobioreactor with a constant light supply [16]. They also fix atmospheric nitrogen and store it in the soil, which promotes plant growth. Moreover, they are natural scavengers [9]. They uptake various toxic pollutants from their surroundings and keep the environment clean. Due to their unique inherent physiological properties, they have been heavily used in numerous applications, including the generation of biofuel and bioplastic, as biofertilizers in sustainable agriculture, and as bioremediation agents for removing toxic pollutants from the environment [9, 12]. This chapter emphasizes the use of cyanobacteria in biofuel and bioplastic production, as well as sustainable agriculture and bioremediation.

2. Cyanobacterial applications

Cyanobacteria have been widely used as model organisms in photosynthesis research [17]. Their remarkable physiological properties have sparked widespread interest in the scientific community. In addition to their involvement in photosynthesis research, they have recently been used in a wide variety of biotechnological applications. Cyanobacteria have been transformed into a biological chassis for numerous industrial and agricultural products, including therapeutic proteins, bioplastics, biofuels, biofertilizers, and remediation agents [9, 12, 18–20] (**Figure 1**).

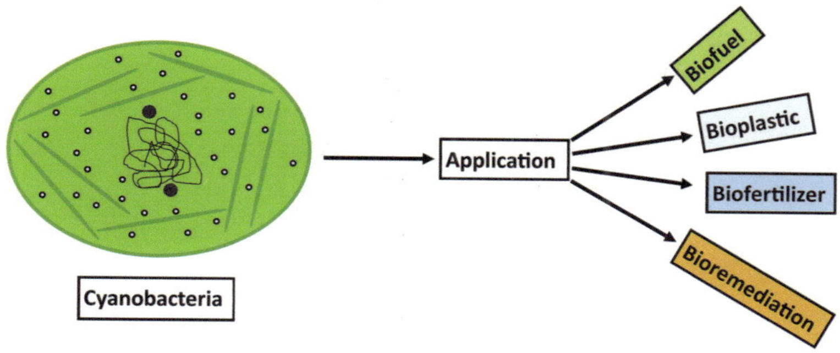

Figure 1.
Overview of cyanobacterial applications.

2.1 Biofuel

Biofuels are renewable green energy sources derived from the biomass of photosynthetic and nonphotosynthetic organisms [21]. Photosynthetic organisms are the largest contributors to biofuel production. Photosynthetic organisms, such as plants, algae, and cyanobacteria, have been used as hosts to produce biofuel as an alternative to fossil fuels [21]. Biofuel has numerous advantages, including lower greenhouse gas emissions, eco-friendliness, and the release of less toxic gases into the environment when burned [22]. Plants, in particular, are heavily used in biofuel production; however, producing biofuel from plants poses numerous challenges, including low yield, time-consuming downstream processing, and high production costs [23]. To alleviate these issues and make biofuel production simpler, more time-efficient, and cost-effective, cyanobacteria have been used as an alternative host [12]. Cyanobacteria's unique physiological properties, such as faster growth, a simple genetic modification process, and economical cultivation methods, make them an ideal host for biofuel production [12]. Several cyanobacterial species, such as *Synechocystis* sp. PCC 6803 [24], *Synechococcus* elongatus PCC 7942, *Synechococcus* sp. PCC 7002 [25, 26], *Nostoc* PCC 7120 [27], *Anabaena* sp. PCC 7120 [28, 29], *Synechococcus* sp. PCC 11901 [30], *Synechococcus elongatus* UTEX 2973 [31, 32], *Synechococcus elongatus* PCC 6301 [33], *Anabaena variabilis* ATCC 29413 [34], *Cyanothece* sp. PCC 7425 [35], and *Cyanothece* sp. ATCC 51142 [36] has been used to produce a wide variety of biofuels, including isoprene, free fatty acids, hydrogen, n-alkanes, ethylene, 2-methyl-1-butanol, 3-methyl-1-butanol, isobutanol, ethanol, limonene, and β-phellandrene (**Table 1**).

Cyanobacteria naturally produce some biofuels, such as ethanol, hydrogen, and free fatty acids, albeit in trace amounts. However, the majority of cyanobacteria lack specific enzymes needed to synthesize biofuels [37]. As a result, cyanobacteria have been genetically and metabolically modified to incorporate heterologous enzymes or entire metabolic pathways specific to biofuel production within the cells [38–40]. The necessary modifications enabled biofuel accumulation in the cells [12]; however, the rates of biofuel production are lower compared to other host species, such as yeast and *E. coli* [41–44]. The limited availability of precursor molecules is the bottleneck contributing to the low rate of biofuel production. To address this issue and increase biofuel production, cyanobacteria have been modified to rewire the carbon flux within the cell, thereby increasing the supply of precursors [45–48]. The

Product	Yield	Strain	References
Isoprene	1.14–3.54 μmol g^{-1} dcw h^{-1}	*Synechocystis* sp. PCC 6803	[49]
	2.80–13.35 μg g^{-1} dcw h^{-1}	*Synechococcus elongatus* UTEX 2973	[50]
	1.7–46.0 μg L^{-1} h^{-1}	*Synechocystis* sp. PCC 6803	[51]
	1.1–5.1 μg L^{-1} h^{-1}	*Synechocystis* sp. PCC 6803	[52]
	4.26 mg L^{-1} h^{-1}	*Synechococcus elongatus* PCC 7942	[53]
Free fatty acids	40% dcw	*Synechocystis* sp. PCC 6803	[54]
	35.8–45.7 mg L^{-1} d^{-1}	*Synechocystis* sp. PCC 6803	[55]
	0.12 mmol L^{-1} $OD_{730}{}^{-1}$	*Synechococcus* sp. PCC 11901	[30]
	0.05 mmol L^{-1} $OD_{730}{}^{-1}$	*Synechococcus* sp. PCC 7002	
Hydrogen	32–546 μmol H2	*Synechocystis* sp. PCC 6803	[56]
	132.3 μmol H2 mg Chl a^{-1} h^{-1}	*Dolichospermum* sp. IPPAS B-1213	[57]
	2–23 nmol H2 mol Chl a^{-1} h^{-1}	*Synechocystis* sp. PCC 6714	[58]
	7–23 nmol H2 mol Chl a^{-1} h^{-1}	*Synechococcus elongatus* PCC 7942 PAMCOD	
n-alkanes	4–27 mg g^{-1} dcw	*Picosynechococcus* sp. strain PCC 7002	[59]
	0.0000007 μg per cell	*Synechocystis* PCC 6803	[60]
	0.15–0.17% dcw	*Synechococcus* sp. PCC 7002	[61]
	0.7–1.5 mol mol Chl a^{-1}	*Synechocystis* PCC 6803	[62]
	0.43 μg mg^{-1} dcw	*Oscillatoria* CCC 305	[63]
	0.13 μg mg^{-1} dcw	*Oscillatoria formosa* BDU 30603	
	300–800 μg g^{-1} dcw	*Synechococcus elongatus* PCC 7942	[64]
	600–1,200 μg g^{-1} dcw	*Anabaena* sp. PCC 7120	
	750–1,600 μg g^{-1} dcw	*Anabaena halophytica*	
Ethylene	2–16 μg mL^{-1} $OD_{730}{}^{-1}$ d^{-1}	*Synechocystis* PCC 6803	[65]
	0–30 μmol L^{-1} h^{-1}	*Synechococcus* PCC 7942	[66]
	6.4–19.2 μg mL^{-1} $OD_{730}{}^{-1}$ d^{-1}	*Synechocystis* PCC 6803	[67]
	338.26 μmol g^{-1} dcw h^{-1}	*Synechococcus elongatus* PCC 11801	[68]
	174–383 μmol mg^{-1} Chl a	*Synechocystis* sp. PCC 6803	[69]
	140 μL L^{-1} $OD_{750}{}^{-1}$ h^{-1}	*Synechococcus elongatus* PCC 7942	[70]
Isobutanol	38 mg L^{-1} $OD_{730}{}^{-1}$	*Synechocystis* sp. PCC 6803	[71]
3-methyl-1-butanol	12 mg L^{-1} $OD_{730}{}^{-1}$		
Isobutanol	27.1–29.8 mg L^{-1} d^{-1}	*Synechocystis* sp. PCC 6803	[72]
3-methyl-1-butanol	7.3–9.5 mg L^{-1} d^{-1}		
Isobutanol	302 mg L^{-1} d^{-1}	*Synechocystis* PCC 6803	[73]
Ethanol	172.6 mg g^{-1} EPS	Atacama cyanobacteria	[74]
	2.11–2.43 g L^{-1}	*Synechocystis* sp. PCC 6803	[75]
	1.2 g L^{-1}	*Synechocystis* PCC 6803	[76]
	248 mg L^{-1} d^{-1}	*Synechocystis* sp. PCC 6803	[77]
	0.25% (v/v)	*Synechococcus* sp. PCC 7002	[78]

Product	Yield	Strain	References
Limonene	8.2 mg L^{-1} d^{-1}	*Synechococcus elongatus* UTEX 2973	[79]
	4.3 mg L^{-1} d^{-1}	*Synechococcus elongatus* PCC 7942	[80]
	0.027 mg L^{-1} d^{-1}	*Synechocystis* PCC 6803	[81]
	0.042 mg L^{-1} d^{-1}	*Synechococcus* PCC 7002	
	4.2 mg L^{-1} d^{-1}	*Synechococcus elongatus* PCC 7942	[82]
	0.95 mg L^{-1} d^{-1}	*Synechocysti* sp. PCC 6803	[83]
β-phellandrene	1.42–5.95 mg g^{-1} dcw	*Synechocysti* sp. PCC 6803	[84]
	3.2 mg g^{-1} dcw	*Synechocysti* sp. PCC 6803	[85]
	1 μg L^{-1} h^{-1}	*Synechocysti* sp. PCC 6803	[86]

dcw: dry cell weight; d: day; L: liter; h: hour; EPS: exopolysaccharide; OD: optical density; Chl: chlorophyll

Table 1.
List of cyanobacteria used in biofuel production.

increased amount of precursors within the cells enhanced biofuel production. The biofuel production rate ranges between 2.4 mg and 2.43 g L^{-1} [12]. Although genetic and metabolic modifications enabled biofuel synthesis in the cell, production rates, cytotoxicity from overproduction of certain biofuels, and low biomass yield remain the biggest challenges in cyanobacterial biofuel production, impeding its commercial success [45–48]. These challenges must be overcome in order to make cyanobacterial biofuel production cost-effective and profitable on a large scale.

2.2 Bioplastic

Every day, humans and other organisms come into contact with plastics, either directly or indirectly. Plastics are used in a variety of applications, ranging from household items to medical devices. Polymers, particularly petrochemicals, are used in the polymerization process to create plastics [87, 88]. Petrochemical-based plastic production is expensive in terms of both money and natural resources. Furthermore, plastics are nonbiodegradable, posing a significant risk of plastic pollution. Plastic pollution is a major global concern because it harms both terrestrial and aquatic organisms [89, 90]. Therefore, it is very important to find an alternative to nondegradable plastics in order to prevent plastic pollution. Bio-based plastics have emerged as a viable alternative to nonbiodegradable plastics in recent decades [91]. Bioplastics are biodegradable and environmentally friendly because they are partially or entirely made up of biological materials. These are divided into different categories based on their biological source, including starch-based, lignin-based, protein-based, chitin-based, and aliphatic-polyester-based bioplastics [92]. Cyanobacteria have recently emerged as a potential candidate for bioplastic production [93, 94] (**Table 2**). They naturally produce a variety of chemicals that can be used to make bioplastics, or they can be genetically modified to produce precursors that can be used in bioplastic production [95, 96]. For example, polyhydroxyalkanoates (PHAs) are lipoidic biopolymers that serve as a major precursor for bio-based plastics production [96]. Polyhydroxybutyrate (PHB), a type of PHA, is a major component of biodegradable plastics produced by various cyanobacterial species [97]. The biosynthetic pathway for PHB production in cyanobacteria involves three enzymes: β-ketothiolase, acetoacetyl-CoA

Product	Yield	Strain	References
PHB	600–4,600 ng mg^{-1} dcw	*Synechocystis* cf. salina 192	[104]
	Up to 19% dcw	*Synechocystis* sp. PCC 6803	[111]
	14–30% dcw	*Nostoc* sp. PCC 7120	[99]
	5.6% dcw	*Synechocystis* sp.	[112]
	26.1% dcw	*Synechococcus* sp.	
	4.6% dcw	*Chlorogloeopsis fritschii*	[113]
	1.1% dcw	*Calothrix* sp.	
	11.4% dcw	*Synechocystis* sp.	
	63–81% dcw	*Synechocystis* sp. PCC 6803	[103]
	0.51% dcw	*Spirulina platensis*	[114]
	8.6% dcw	*Nostoc muscorum*	
	0.06 g $g_{biomass}^{-1}$ d^{-1}	*Anabaena* sp.	[115]
	13–16.4% dcw	*Synechocystis* sp. PCC 6714	[102]
	5.04% dcw	*Synechocystis* sp.	[116]
	8% dcw	*Synechocystis* PCC 6803	[117]
	17.31% dcw	*Oscillatoria salina*	[118]
	15.22% dcw	*Anabaena cylindrica*	
	14.95% dcw	*Nostoc linckia*	
	26–35% dcw	*Synechocystis* sp. PCC 6803	[101]
	28.2–39% dcw	*Arthrospira platensis* NLHT3	[119]
	Up to 85% dcw	*Aulosira fertilissima*	[105]
	34.9% dcw	*Synechococcus elongatus* PCC 7942	[109]
	3–11% dcw	*Synechocystis* cf. *salina Wislouch*	[108]
	Up to 16% dcw	*Synechocystis* sp. PCC 6714	[107]
	Up to 41.72% dcw	*Synechocystis* sp. PCC 6803	[110]
	Up to 3.1% dcw	*Synechocystis* sp. PCC 6803	[106]

dcw: dry cell weight; d: day

Table 2.
List of cyanobacteria used in polyhydroxybutyrate (PHB) production.

dehydrogenase, and PHA synthase. Acetyl–CoA is converted to acetoacetyl-CoA by β-ketothiolase, which is then converted to R-3-hydroxybutyryl–CoA by acetoacetyl-CoA dehydrogenase. PHA synthase then polymerizes this to PHB [92]. However, the rate of PHB production in cyanobacteria, especially in photoautotrophic conditions, is much lower compared to other host species [98, 99]. Recent studies show that cyanobacteria can accumulate PHB up to 85% of their dry cell weight (dcw) in nutrient-deficient and supplemented conditions [100]. For example, growing cyanobacteria in nitrogen and phosphorus-depleted conditions shows enhanced PHB accumulation [101–104]. In addition to nutrient-deprived conditions, growth media supplemented with acetate, citrate, glucose, and valerate enhance PHB production in cyanobacteria by up to 85% dcw [103, 105]. Moreover, genetic and metabolic modifications, along with carbon flux rewiring, have shown increased PHB production in various cyanobacterial species [99, 103, 106–110]. Although bio-based plastics are more cost-effective and environmentally friendly, their production is far below expected demand. Implementing the aforementioned methods, as well as optimizing environmental and growth conditions, could alleviate the low-yield issue in cyanobacteria.

2.3 Biofertilizer

Agriculture is the only option for feeding the world's growing population. However, modern agriculture heavily relies on chemical fertilizers and pesticides to increase the quantity of agricultural products, which are not eco-friendly as they have negative impacts on the environment and human health [120, 121]. Furthermore, the production cost is high due to the rising prices of chemical fertilizers and pesticides, which have an indirect effect on the global economy. As a result, it is preferable to seek alternatives to this chemical-based farming method. Many agricultural farms are now practicing sustainable farming to ensure a safe environment and healthy food. Sustainable farming is a new method of cultivating crops and producing food that eliminates the need for chemical fertilizers. It has numerous advantages, including producing healthy and safe food at a lower cost, preserving the natural microbial population in the soil, and being eco-friendly without endangering the environment [122]. Cyanobacteria have been considered a potential candidate for sustainable farming because of their ability to increase soil fertility, protect crop plants from various pathogens, prevent heavy metals from entering the food chain, promote plant growth hormone production, and facilitate mineral solubilization and uptake from soils to plants [123–127] (**Table 3**) (**Figure 2**).

Cyanobacterial species such as *Nostoc*, *Calothrix*, *Oscillatoria* sp., *Rivularia*, *Westiella*, *Fischerella*, *Gloeotirchia*, *Scytonema*, *Tolypothrix*, *Anabaena*, and *Aulosira* have specialized cells called heterocysts that contain nitrogenase, an enzyme involved in nitrogen fixation [141–143]. However, nitrogenase is oxygen sensitive and becomes inactive when exposed to it [144]. As a result, cyanobacteria have evolved various mechanisms to avoid inactivation, including creating an oxygen-deprived environment, performing nitrogen fixation at specific times and periods, developing oxygen-tolerant nitrogenase, and overexpressing nitrogenase in the cells [138, 145–149]. Nitrogen fixation rates vary among cyanobacterial species and are also affected by environmental conditions. The average rate of nitrogen fixation by cyanobacteria is 0–1.0 nmol h^{-1} L^{-1} [135]. Cyanobacterial nitrogen fixation in agricultural fields, particularly in rice fields, is very high. In addition to their nitrogen-fixing ability, some cyanobacterial species, such as Anabaena, *Phormidium*, *Calothrix*, *Trichormus*, *Chlorogloeopsis*, *Spirulina*, *Cylindrospermum*, *Glactothece*, *Nostoc*, *Plectonema*, and *Synechocystis* secrete a variety of chemicals into their surroundings, including abscisic acids, auxin, amino acids, and cytokinins, which promote plant growth and development [9, 143] (**Table 4**). These chemicals play numerous roles in plant growth and development [150–153]. Using cyanobacteria in agriculture has numerous benefits, including enriching the soil with various essential components that promote plant growth and development while also protecting the soil from erosion. Cyanobacteria not only promote plant growth, but they also protect them from pathogens. They produce a wide range of bioactive compounds that have antibacterial, antifungal, and antiviral properties [154, 155]. For example, cyanobacterial crude extracts have been shown in numerous studies to

Nitrogen fixation rate	Strain	References
0.67 nmol N 0.1 g^{-1} dcw h^{-1}	*Nostoc commune*	[128]
0.08 μmol N g^{-1} Chl s^{-1}	*Nostoc punctiforme*	[129]
0.25 μmol N g^{-1} Chl s^{-1}	G23 (*Nostoc* NIES 211)	
0.26 μmol N g^{-1} Chl s^{-1}	G42 (*Anabaena cylindrical*)	
0.027–2.85 nmol N min^{-1} g^{-1} FW	*Nostoc punctiforme*	[130]
8.9–83.9 mg N L^{-1} d^{-1}	*Nostoc* sp. PCC 7120	[131]
31.4 nmol N $μg^{-1}$ Chl d^{-1}	*Oscillatoria* sp.	[132]
0.12–0.17 fmol N $heterocyst^{-1}$ h^{-1}	*Calothrix rhizosoleniae* and *Richelia intracellularis*	[133]
10.04	*Schizothrix*	[134]
16.6 nmol N mg^{-1} chl h^{-1}	*Rivularia*	
0–0.85 nmol N L^{-1} h^{-1}	*Nodularia* and *Dolichospermum*	[135]
68–301.5 nmol N mg^{-1} Chl h^{-1}	*Thermoleptolyngbya* sp. O-77	[136]
0.89 nmol N mg^{-1} dcw h^{-1}	*Trichormus variabilis* K-31	[137]
0.70 nmol N mg^{-1} dcw h^{-1}	*Nostoc* sp. J-14	
0.55 nmol N mg^{-1} dcw h^{-1}	*Tolypothrix tenuis* J-1	
0.34 nmol N mg^{-1} dcw h^{-1}	*Cylindrospermum badium* J-8	
22.0 μmol mg^{-1} Chl h^{-1}	*Fischerella* sp.	[138]
1.36 nmol N $μg^{-1}$ Chl h^{-1}	*Hapalosiphon*	[139]
0.23 nmol N $μg^{-1}$ Chl h^{-1}	*Chlorogloeopsis*-354	
0.64 nmol N $μg^{-1}$ Chl h^{-1}	*Westiellopsis*	
50.34 U mg^{-1} dcw	*Anabaena*	[140]
75.33 U mg^{-1} dcw	*Desmonostoc muscorum*	

FW: fresh weight; dcw: dry cell weight; Chl: chlorophyll

Table 3.
List of cyanobacteria performing nitrogen fixation.

protect plants from pathogens [156–159]. Furthermore, cyanobacteria can improve aeration and fertility of various types of soils, including salty soils [160–163]. Although cyanobacteria offer numerous benefits for sustainable agriculture, they should be used with caution as they produce cyanotoxins, which are harmful to both aquatic animals and humans [164, 165].

2.4 Bioremediation

Cyanobacteria are extremely beneficial for sustainable agriculture because they provide various nutrients, growth factors, and pathogen protection to plants [125]. They not only protect and promote plant growth and development, but they are also excellent candidates for bioremediation [171–173]. They can scavenge many toxic substances from the environment, including heavy metals, toxic industrial effluents, and other hazardous substances [174–177] (**Table 5**) (**Figure 3**). Cyanobacteria perform bioremediation through a variety of processes, including biosorption, biotransformation, and bioaccumulation [173]. Biosorption is a process where

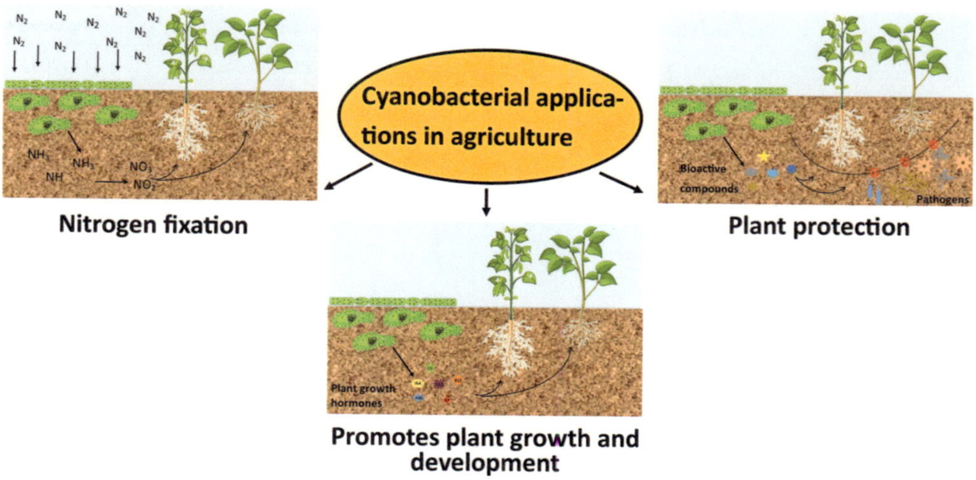

Figure 2.
Cyanobacteria applications in sustainable agriculture. Cyanobacterial applications include fixing atmospheric nitrogen, enhancing soil fertility and aeration, providing plant protection, and promoting plant growth and development.

Product	Yield	Strain	References
Auxin	5.5 μg g^{-1} FW	*Planktothricoides raciborskii*	[166]
	0.9 μg mL^{-1}	*Leptolyngbya* SM-13	[167]
	2.5 μg mL^{-1}	*Chroococcidiopsis* SM-04	
	4.0 μg mL^{-1}	*Phormidium* SM-14	
	3.6 μg mL^{-1}	*Synechocystis* SM-10	
	9.4 pmol g^{-1} FW	*Nostoc microscopicum*	[168]
	192.9 pmol g^{-1} FW	*Chroococcus minutus*	
	250.1 pmol g^{-1} FW	*Phormidium animale*	
Cytokinin	0.0063 μg mL^{-1}	*Nostoc*	[169]
	0.0059 μg mL^{-1}	*Anabaena*	
	0.0026 μg mL^{-1}	*Synechococcus* sp.	
Abscisic acid	1.0 μg L^{-1}	*Synechococcus leopoliensis*, *Nostoc muscorum*, *Trichormus variabilis*	[170]

FW: fresh weight

Table 4.
List of cyanobacteria produce plant hormones.

cyanobacteria passively bind heavy metals through extracellular polysaccharides on their cell walls. Cyanobacterial species from the genera *Synechococcus*, *Nostoc*, *Spirulina*, *Oscillatoria*, *Microcystis*, and *Phormidium* have been found to be effective at accumulating heavy metals from aqueous solutions, including Ni, Cu, Pb, Cd, and Zn [178–184]. The rate of biosorption in cyanobacteria ranges between 2 and 515 mgg^{-1} dcw [185].

Rate	Name of chemical/metals	Strain	References
18.19 mg g^{-1} biomass	Nickel (Ni^{2+})	*Chroococcidiopsis* sp.	[203]
13.40 mg g^{-1} biomass		*Arthrospira platensis*	[204]
38.0–63.80%		*Nostoc* sp.	[205]
98.03 mg g^{-1} biomass	Zinc (Zn^{2+})	*Fischerella ambigua* ISC67	[206]
89–95%		*Microcystis aeruginosa*	[184]
1.80 mg g^{-1} biomass		*Nostoc commune* R123	[183]
1.70 mg g^{-1} biomass		*Nostoc commune* AB55	
2.0 mg g^{-1} biomass		*Nostoc commune* NR64	
82–87%		*Nostoc* sp.	[205]
80–100%		*Nostoc linckia*	[207]
92.27–98.15%	Lead (Pb^{2+})	*Nostoc* sp.	[205]
89–95%		*Microcystis aeruginosa*	[184]
3.90 mg g^{-1} biomass		*Nostoc commune* R123	[183]
2.40 mg g^{-1} biomass		*Nostoc commune* AB55	
1.80 mg g^{-1} biomass		*Nostoc commune* NR64	
86%		*Microcystis aeruginosa*	[208]
90–100%		*Microcystis aeruginosa*	[209]
95.14–96.0%	Copper (Cu^{2+})	*Nostoc* sp.	[205]
Up to 45%		*Cyanothece* sp. CE4	[210]
65–90%		*Nostoc linckia*	[207]
0.0 mg g^{-1} biomass		*Nostoc commune* R123	[183]
4.10 mg g^{-1} biomass		*Nostoc commune* AB55	
1.20 mg g^{-1} biomass		*Nostoc commune* NR64	
92%	Cadmium (Cd^{2+})	*Microcystis aeruginosa*	[208]
4.0–17.7 mg g^{-1} biomass		*Synechocystis* sp. PCC6803	[211]
93.4%		*Nostoc muscorum*	[212]
89.13%		*Trichormus variabilis*	[212]
69.5–98.4%		*Microcystis aeruginosa*	[213]
80-100%		*Microcystis aeruginosa*	[209]
91.0%		*Nostoc* sp.	[205]

Rate	Name of chemical/metals	Strain	References
60.7–92.4%	Aluminum (Al^{3+})	*Microcystis aeruginosa*	[213]
73–100%	Iron (Fe^{2+})	*Nostoc linckia*	[207]
84.2–91.5 mg g^{-1} biomass	Cerium (Ce^{3+}), Neodymium (Nd^{3+}), Terbium (Tb^{3+}), Lanthanum (La^{3+})	*Nostoc* sp. 20.02	[214]
69.5–83.4 mg g^{-1} biomass		*Synechococcus elongatus* UTEX 2973	
68.6–83.5 mg g^{-1} biomass		*Calothrix brevissima* SAG 34.79	
44.7–70.6 mg g^{-1} biomass		*Desmonostoc muscorum* 90.03	
47.2–67.1 mg g^{-1} biomass		*Komarekiella* sp. 89.12	
63%	Chromium (Cr^{6+})	*Limnococcus* sp.	[192]
96%		*Spirulina fusiformis*	
92.91%	Uranium (U^{6+})	*Anagnostidinema amphibium*	[215]
25.4–56.27 mg g^{-1} biomass	Cesium (Cs)	*Synechococcus* PCC 7002	[216]
62–90%	Strontium (^{90}Sr)	*Arthrospira platensis* CNMN-CB-02	[217]
15–99.7%	Neptunium (^{237}Np)		
58%	Cesium (^{137}Cs)		
3–8%	Uranium (^{233}U)		
67–98.7	Plutonium (^{239}Pu)		
77–99.4%	Americium (^{241}Am)		

Table 5.
List of cyanobacteria used in bioremediation.

Furthermore, some cyanobacteria have been shown to be effective in removing a variety of toxic metals from wastewater and industrial effluents through bioaccumulation [186–189]. Cyanobacteria transport heavy metals from their surroundings to their cells using various passive and active transporters, as well as ion channels in the membrane [173]. The rate of bioaccumulation varies by species and metal type. For example, the bioaccumulation efficiency of *Nostoc* sp. for Cd^{2+}, Hg^{2+}, and Pb^{2+} is 29%, 34%, and 34%, respectively, whereas the efficiency rates of *Synechococcus* sp. PCC 7942 for the same metals are 34%, 50%, and 45%, respectively [190]. Similarly, *Limnococcus* sp. and *Spirulina fusiformis* have a bioaccumulation efficiency of 63% and 96% for Cr^{6+}, respectively [191, 192]. Cyanobacteria not only accumulate various toxic metals from their surroundings, but they also detoxify and convert them into less toxic forms within the cell. They biotransform toxic metals using repeated oxidation and reduction reactions.

Current agricultural methods heavily rely on pesticides and herbicides to protect crop plants from pathogens and wild flora. As a result, nearby water bodies become contaminated with pesticides and herbicides. Cyanobacteria have shown promise in degrading these chemicals from polluted water. Cyanobacteria such as *Nostoc muscorum*, *Microcystis aeruginosa*, *Anabaena,* and *Fischerella sp.* have been

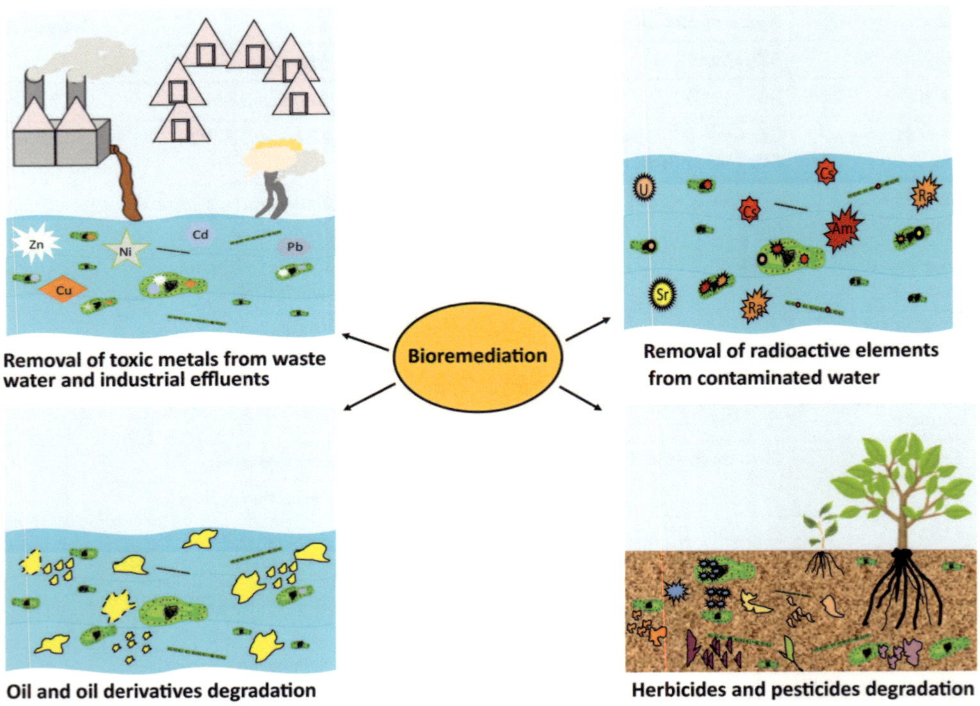

Figure 3.
Overview of cyanobacterial applications in bioremediation. Cyanobacterial applications include the removal of toxic metals from waste water and industrial effluents, the removal of radioactive elements from contaminated water, the degradation of oil and oil derivatives from water, and the degradation of herbicides and pesticides from contaminated soil.

demonstrated to successfully degrade pesticides from agricultural fields and contaminated water bodies [193–196]. Similarly, numerous studies have utilized cyanobacteria as hosts to degrade herbicides from polluted soil. Cyanobacteria, such as *Anabaena*, *Microcystis*, *Nostoc*, *Synechococcus elongatus*, and *Spirulina,* have shown promising results in breaking down herbicides from their surrounding contaminated soil [197–202]. Although they are naturally potent at removing toxic chemicals from the environment, the rate of removal is not high and can be improved through optimization. Thus, cyanobacteria could be a viable candidate for large-scale bioremediation in the near future.

3. Conclusion

Cyanobacteria have great potential for biomanufacturing, as well as for sustainable agriculture and bioremediation. Numerous studies using cyanobacteria as hosts have shown promising results in terms of bioproduct generation, toxic pollutant and metal removal from the environment, and plant growth. However, product yield, contamination, and health risks remain hurdles to large-scale applications. Genetically modified strains are capable of enhancing bioproduct formation, toxic metal removal, and nitrogen fixation. However, the use of genetically engineered strains is associated with environmental risks. As a result, the adaptive laboratory evolution method can be used as an alternative to genetic modification for developing highly efficient strains. Thus, the use of highly efficient strains, combined with optimized downstream processing, could alleviate current challenges and transform cyanobacteria into a cost-effective biological chassis for a variety of applications.

Author details

Bharat Kumar Majhi
University of California, Department of Plant and Microbial Biology, Berkeley, CA, USA

*Address all correspondence to: bkm6728@berkeley.edu

References

[1] Khan S, Ullah MW, Siddique R, Nabi G, Manan S, Yousaf M, Hou H. Role of recombinant DNA technology to improve life. International Journal of Genomics. 2016;**2016**(1):2405954. DOI: 10.1155/2016/2405954

[2] Karataş P, Ayaz F. Synthetic biology and application areas. Discover Biotechnology. 2025;**2**(1):3. DOI: 10.1007/s44340-025-00010-5

[3] Iram A, Dong Y, Ignea C. Synthetic biology advances towards a bio-based society in the era of artificial intelligence. Current Opinion in Biotechnology. 2024;**87**:103143. DOI: 10.1016/j.copbio.2024.103143

[4] Dumont J, Euwart D, Mei B, Estes S, Kshirsagar R. Human cell lines for biopharmaceutical manufacturing: History, status, and future perspectives. Critical Reviews in Biotechnology. 2016;**36**(6):1110–1122. DOI: 10.3109/07388551.2015.1084266

[5] Serra LA, Trichez D, Carneiro CVGC, Ferreira LMM, Franco PF, and Almeida JRM. Microbial production of biobased chemicals. In: Lignocellulose Bioconversion Through White Biotechnology (Wiley); 2022. p. 136–176. DOI: 10.1002/9781119735984

[6] Syrpas M, and Venskutonis PR. Chapter 6 - Algae for the production of bio-based products. In: Galanakis CM, editor. Biobased Products and Industries. Amsterdam, Netherlands: Elsevier; 2020. p. 203–243. DOI: 10.1016/B978-0-12-818493-6.00006-3

[7] Lopez L, Alagna F, Bianco L, De Bari I, Fasano C, Panara F, and Perrella G. Chapter 3 - Plants: A sustainable platform for second-generation biofuels and biobased chemicals. In: Sahay S, editor. Handbook of Biofuels. London, UK: Academic Press; 2022. p. 47–72. DOI: 10.1016/B978-0-12-822810-4.00003-8

[8] Melis A, Martinez DAH, Betterle N. Perspectives of cyanobacterial cell factories. Photosynthesis Research. 2024; **162**(2):459–471. DOI: 10.1007/s11120-023-01056-4

[9] Singh JS, Kumar A, Rai AN, Singh DP. Cyanobacteria: A precious bio-resource in agriculture, ecosystem, and environmental sustainability. Frontiers in Microbiology. 2016;7. DOI: 10.3389/fmicb.2016.00529

[10] Yang Y, Chaffin TA, Ahkami AH, Blumwald E, Stewart CN. Plant synthetic biology innovations for biofuels and bioproducts. Trends in Biotechnology. 2022;**40**(12):1454–1468. DOI: 10.1016/j.tibtech.2022.09.007

[11] Barbosa MJ, Janssen M, Südfeld C, D'Adamo S, Wijffels RH. Hypes, hopes, and the way forward for microalgal biotechnology. Trends in Biotechnology. 2023;**41**(3):452–471. DOI: 10.1016/j.tibtech.2022.12.017

[12] Majhi BK. Cyanobacteria: Photosynthetic cell factories for biofuel production. Journal of Bioresources and Bioproducts. 2025;**10**(2):128–144. DOI: 10.1016/j.jobab.2024.10.001

[13] Soo RM, Hemp J, Parks DH, Fischer WW, Hugenholtz P. On the origins of oxygenic photosynthesis and

aerobic respiration in cyanobacteria. Science. 2017;**355**(6332):1436–1440. DOI: 10.1126/science.aal3794

[14] Hitchcock A, Hunter CN, Canniffe DP. Progress and challenges in engineering cyanobacteria as chassis for light-driven biotechnology. Microbial Biotechnology. 2020;**13** (2):363–367. DOI: 10.1111/1751-7915.13526

[15] Arias DM, García J, Uggetti E. Production of polymers by cyanobacteria grown in wastewater: Current status, challenges and future perspectives. New Biotechnology. 2020;**55**:46–57. DOI: 10.1016/j.nbt.2019.09.001

[16] Novoveská L, Nielsen SL, Eroldoğan OT, Haznedaroglu BZ, Rinkevich B, Fazi S, Robbens J, Vasquez M, Einarsson H. Overview and challenges of large-scale cultivation of photosynthetic microalgae and cyanobacteria. Marine Drugs. 2023;**21** (8):445. DOI: 10.3390/md21080445

[17] Allaf MM, Peerhossaini H. Cyanobacteria: Model microorganisms and beyond. Microorganisms. 2022;**10**(4). DOI: 10.3390/microorganisms10040696

[18] Rueda E, Senatore V, Zarra T, Naddeo V, García J, Garfí M. Life cycle assessment and economic analysis of bioplastics production from cyanobacteria. Sustainable Materials and Technologies. 2023;**35**:e00579. DOI: 10.1016/j.susmat.2023.e00579

[19] Majhi BK, Melis A. Recombinant protein synthesis and isolation of human interferon alpha-2 in cyanobacteria. Bioresource Technology. 2024;**400**:130664. DOI: 10.1016/j.biortech.2024.130664

[20] Majhi BK, Melis A. Recombinant protein synthesis and isolation of human interferon alpha-2 in cyanobacteria. Cell Signaling. 2025;**3**:23–26. DOI: 10.46439/signaling.3.051

[21] El-Araby R. Biofuel production: Exploring renewable energy solutions for a greener future. Biotechnology for Biofuels and Bioproducts. 2024;**17**(1):129. DOI: 10.1186/s13068-024-02571-9

[22] Hill J, Nelson E, Tilman D, Polasky S, Tiffany D. Environmental, economic, and energetic costs and benefits of biodiesel and ethanol biofuels. Proceedings of the National Academy of Sciences. 2006;**103** (30):11206–11210. DOI: 10.1073/pnas.0604600103

[23] Russo C, Cirillo V, Pollaro N, Terribile F, Chiodini A, Maggio A. The global energy challenge: Second-generation feedstocks on marginal lands for a sustainable biofuel production. Chemical and Biological Technologies in Agriculture. 2025;**12** (1):10. DOI: 10.1186/s40538-025-00729-7

[24] Rodrigues JS, Lindberg P. Metabolic engineering of Synechocystis sp. PCC 6803 for improved bisabolene production. Metabolic Engineering Communications. 2021;**12**:e00159. DOI: 10.1016/j.mec.2020.e00159

[25] Gonzales JN, Treece TR, Mayfield SP, Simkovsky R, Atsumi S. Utilization of lignocellulosic hydrolysates for photomixotrophic chemical production in Synechococcus elongatus PCC 7942. Communications Biology. 2023;**6**(1):1022. DOI: 10.1038/s42003-023-05394-w

[26] Srivastava V, Sarnaik AP, Wangikar PP. Metabolic engineering of

rapidly growing Synechococcus elongatus strains for phototrophic production of alkanes. Biotechnology Progress. 2025;**41**(1):e3509. DOI: 10.1002/btpr.3509

[27] Roumezi B, Avilan L, Risoul V, Brugna M, Rabouille S, Latifi A. Overproduction of the Flv3B flavodiiron, enhances the photobiological hydrogen production by the nitrogen-fixing cyanobacterium Nostoc PCC 7120. Microbial Cell Factories. 2020;**19**(1):65. DOI: 10.1186/s12934-020-01320-5

[28] Singh R, Kumar A, Sharma YC. Biodiesel synthesis from microalgae (Anabaena PCC 7120) by using barium titanium oxide (Ba2TiO4) solid base catalyst. Bioresource Technology. 2019;**287**:121357. DOI: 10.1016/j.biortech.2019.121357

[29] Pandit S, Sharma M, Banerjee S, Nayak BK, Das D, Khilari S, Prasad R. Pretreatment of cyanobacterial biomass for the production of biofuel in microbial fuel cells. Bioresource Technology. 2023;**370**:128505. DOI: 10.1016/j.biortech.2022.128505

[30] Włodarczyk A, Selão TT, Norling B, Nixon PJ. Newly discovered Synechococcus sp. PCC 11901 is a robust cyanobacterial strain for high biomass production. Communications Biology. 2020;**3**(1):215. DOI: 10.1038/s42003-020-0910-8

[31] Li S, Sun T, Chen L, Zhang W. Light and carbon dioxide-driven synthesis of high-density fuel in Synechococcus elongates UTEX 2973], Sheng wu gong cheng xue bao. Chinese Journal of Biotechnology. 2020;**36**(10):2126–2138. DOI: 10.13345/j.cjb.200090

[32] Rautela A, Yadav I, Gangwar A, Chatterjee R, Kumar S. Photosynthetic production of α-farnesene by engineered Synechococcus elongatus UTEX 2973 from carbon dioxide. Bioresource Technology. 2024;**396**:130432. DOI: 10.1016/j.biortech.2024.130432

[33] Liang F, Englund E, Lindberg P, Lindblad P. Engineered cyanobacteria with enhanced growth show increased ethanol production and higher biofuel to biomass ratio. Metabolic Engineering. 2018;**46**:51–59. DOI: 10.1016/j.ymben.2018.02.006

[34] Salleh SF, Kamaruddin A, Uzir MH, Mohamed AR, Shamsuddin AH. Kinetic modeling of hydrogen production rate by photoautotrophic cyanobacterium A. variabilis ATCC 29413 as a function of both CO(2) concentration and oxygen production rate. Preparative Biochemistry & Biotechnology. 2017;**47**(2):111–115. DOI: 10.1080/10826068.2016.1181085

[35] Chenebault C, Diaz-Santos E, Kammerscheit X, Görgen S, Ilioaia C, Streckaite S, Gall A, Robert B, Marcon E, Buisson D-A, Benzerara K, Sassi J-F, Cassier-Chauvat C, Chauvat F. A genetic toolbox for the new model cyanobacterium cyanothece PCC 7425: A case study for the photosynthetic production of limonene. Frontiers in Microbiology. 2020;**11**:2020. DOI: 10.3389/fmicb.2020.586601

[36] Valvassore M, Freitas H, Andrade C, Costa C. Improving feeding profile strategy for hydrogen production by Cyanothece sp. ATCC 51142 using meta-heuristic methods. Chemical Engineering Communications. 2021;**210**:1–15. DOI: 10.1080/00986445.2021.1986701

[37] Velmurugan R, Incharoensakdi A. Metabolic transformation of

cyanobacteria for biofuel production. Chemosphere. 2022;**299**:134342. DOI: 10.1016/j.chemosphere.2022.134342

[38] Santos-Merino M, Singh AK, Ducat DC. New applications of synthetic biology tools for cyanobacterial metabolic engineering. Frontiers in Bioengineering and Biotechnology. 2019;7. DOI: 10.3389/fbioe.2019.00033

[39] Taylor GM, Hitchcock A, Heap JT. Combinatorial assembly platform enabling engineering of genetically stable metabolic pathways in cyanobacteria. Nucleic Acids Research. 2021;**49**(21):e123–e123. DOI: 10.1093/nar/gkab791

[40] Liu X, Tang K, Hu J. Application of Cyanobacteria as chassis cells in synthetic biology. Microorganisms. 2024;**12**(7):1375

[41] Liang L, Liu R, Freed EF, Eckert CA. Synthetic biology and metabolic engineering employing escherichia coli for C2–C6 bioalcohol production. Frontiers in Bioengineering and Biotechnology. 2020;**8**. DOI: 10.3389/fbioe.2020.00710

[42] Liu H, Zhou P, Qi M, Guo L, Gao C, Hu G, Song W, Wu J, Chen X, Chen J, Chen W, Liu L. Enhancing biofuels production by engineering the actin cytoskeleton in Saccharomyces cerevisiae. Nature Communications. 2022;**13**(1):1886. DOI: 10.1038/s41467-022-29560-6

[43] Liu Z, Wang J, Nielsen J. Yeast synthetic biology advances biofuel production. Current Opinion in Microbiology. 2022;**65**:33–39. DOI: 10.1016/j.mib.2021.10.010

[44] Wang C, Chang D, Zhang Q, Yu Z. Enhanced bioethanol production by evolved Escherichia coli LGE2-H in a microbial electrolysis cell system. Bioresources and Bioprocessing. 2024;**11**(1):4. DOI: 10.1186/s40643-023-00717-5

[45] Wang F, Gao Y, Yang G. Recent advances in synthetic biology of cyanobacteria for improved chemicals production. Bioengineered. 2020;**11**(1):1208–1220. DOI: 10.1080/21655979.2020.1837458

[46] Treece TR, Gonzales JN, Pressley JR, Atsumi S. Synthetic biology approaches for improving chemical production in cyanobacteria. Frontiers in Bioengineering and Biotechnology. 2022;**10**. DOI: 10.3389/fbioe.2022.869195

[47] Gao E-B, Wu J, Ye P, Qiu H, Chen H, Fang Z. Rewiring carbon flow in Synechocystis PCC 6803 for a high rate of CO2-to-ethanol under an atmospheric environment. Frontiers in Microbiology. 2023;**14**. DOI: 10.3389/fmicb.2023.1211004

[48] Mescouto VAD, Ferreira LDC, Paiva RDJ, Oliveira DTD, de Oliveira MS, Filho GNDR, Luque R, Noronha RCR, Nascimento LASD. Review of recent advances in improvement strategies for biofuels production from cyanobacteria. Heliyon. 2024;**10**(22):e40293. DOI: 10.1016/j.heliyon.2024.e40293

[49] Rodrigues JS, Kovács L, Lukeš M, Höper R, Steuer R, Červený J, Lindberg P, Zavřel T. Characterizing isoprene production in cyanobacteria – Insights into the effects of light, temperature, and isoprene on Synechocystis spPCC 6803. Bioresource Technology. 2023;**380**:129068. DOI: 10.1016/j.biortech.2023.129068

[50] Yadav I, Rautela A, Gangwar A, Kesari V, Padhi AK, Kumar S. Geranyl diphosphate synthase (CrtE) inhibition using alendronate enhances isoprene production in recombinant Synechococcus elongatus UTEX 2973: A step towards isoprene biorefinery. Fermentation. 2023;**9**(3):217

[51] Chaves JE, Melis A. Engineering isoprene synthesis in cyanobacteria. FEBS Letters. 2018;**592**(12):2059–2069. DOI: 10.1002/1873-3468.13052

[52] Chaves JE, Romero PR, Kirst H, Melis A. Role of isopentenyl-diphosphate isomerase in heterologous cyanobacterial (Synechocystis) isoprene production. Photosynthesis Research. 2016;**130**(1):517–527. DOI: 10.1007/s11120-016-0293-3

[53] Gao X, Gao F, Liu D, Zhang H, Nie X, Yang C. Engineering the methylerythritol phosphate pathway in cyanobacteria for photosynthetic isoprene production from CO2. Energy & Environmental Science. 2016;**9**(4):1400–1411. DOI: 10.1039/C5EE03102H

[54] Eungrasamee K, Lindblad P, Jantaro S. Enhanced productivity of extracellular free fatty acids by gene disruptions of acyl-ACP synthetase and S-layer protein in Synechocystis sp. PCC 6803. Biotechnology for Biofuels and Bioproducts. 2022;**15**(1):99. DOI: 10.1186/s13068-022-02197-9

[55] Eungrasamee K, Incharoensakdi A, Lindblad P, Jantaro S. Synechocystis sp. PCC 6803 overexpressing genes involved in CBB cycle and free fatty acid cycling enhances the significant levels of intracellular lipids and secreted free fatty acids. Scientific Reports. 2020;**10**(1):4515. DOI: 10.1038/s41598-020-61100-4

[56] Appel J, Hueren V, Boehm M, Gutekunst K. Cyanobacterial in vivo solar hydrogen production using a photosystem I–hydrogenase (PsaD-HoxYH) fusion complex. Nature Energy. 2020;**5**(6):458–467. DOI: 10.1038/s41560-020-0609-6

[57] Bozieva AM, Khasimov MK, Rao MS, Sinetova MA, Voloshin RA, Dunikov DO, Tsygankov AA, Leong YK, Chang J-S, Allakhverdiev SI, Bruce BD. Optimizing cyanobacterial hydrogen production: Metabolic and genetic strategies with glycerol supplementation. Frontiers in Energy Research. 2025;**13**. DOI: 10.3389/fenrg.2025.1547215

[58] Broussos P-I, Romanos GE, Stamatakis K. Salt and heat stress enhances hydrogen production in cyanobacteria. Photosynthesis Research. 2024;**161**(1):117–125. DOI: 10.1007/s11120-024-01098-2

[59] Yñigez-Gutierrez AE, Conley E, Thomas MG, Pfleger BF. Uncovering the substrate of olefin synthase loading domains in cyanobacteria Picosynechococcus sp. strain PCC 7002. RSC Chemical Biology. 2025;**6**(2):307–316. DOI: 10.1039/D4CB00234B

[60] Miao R, Légeret B, Cuine S, Burlacot A, Lindblad P, Li-Beisson Y, Beisson F, Peltier G. Absence of alka(e)nes triggers profound remodeling of glycerolipid and carotenoid composition in cyanobacteria membrane. Plant Physiology. 2024;**196**(1):397–408. DOI: 10.1093/plphys/kiae319

[61] Knoot CJ, Pakrasi HB. Diverse hydrocarbon biosynthetic enzymes can substitute for olefin synthase in the

cyanobacterium Synechococcus sp. PCC 7002. Scientific Reports. 2019;**9**(1):1360. DOI: 10.1038/s41598-018-38124-y

[62] Berla BM, Saha R, Maranas CD, Pakrasi HB. Cyanobacterial alkanes modulate photosynthetic cyclic electron flow to assist growth under cold stress. Scientific Reports. 2015;**5**(1):14894. DOI: 10.1038/srep14894

[63] Shakeel T, Fatma Z, Fatma T, Yazdani SS. Heterogeneity of alkane chain length in freshwater and marine cyanobacteria. Frontiers in Bioengineering and Biotechnology. 2015;**3**. DOI: 10.3389/fbioe.2015.00034

[64] Kageyama H, Waditee-Sirisattha R, Sirisattha S, Tanaka Y, Mahakhant A, Takabe T. Improved alkane production in nitrogen-fixing and halotolerant cyanobacteria via abiotic stresses and genetic manipulation of alkane synthetic genes. Current Microbiology. 2015;**71**(1):115–120. DOI: 10.1007/s00284-015-0833-7

[65] Gao X, Liu X, Jing X, Lindblad P. Anchoring the ethylene-forming enzyme to photosystem-related proteins to improve ethylene production in engineered Synechocystis PCC 6803. Bioresource Technology Reports. 2022;**19**:101178. DOI: 10.1016/j.biteb.2022.101178

[66] Wang B, Xu Y, Wang X, Yuan JS, Johnson CH, Young JD, Yu J. A guanidine-degrading enzyme controls genomic stability of ethylene-producing cyanobacteria. Nature Communications. 2021;**12**(1):5150. DOI: 10.1038/s41467-021-25369-x

[67] Durall C, Lindberg P, Yu J, Lindblad P. Increased ethylene production by overexpressing phosphoenolpyruvate carboxylase in the cyanobacterium Synechocystis PCC 6803. Biotechnology for Biofuels. 2020;**13**(1):16. DOI: 10.1186/s13068-020-1653-y

[68] Sengupta A, Pritam P, Jaiswal D, Bandyopadhyay A, Pakrasi HB, Wangikar PP. Photosynthetic co-production of succinate and ethylene in a fast-growing cyanobacterium, synechococcus elongatus PCC 11801. Metabolites. 2020;**10**(6):250

[69] Vajravel S, Sirin S, Kosourov S, Allahverdiyeva Y. Towards sustainable ethylene production with cyanobacterial artificial biofilms. Green Chemistry. 2020;**22**(19):6404–6414. DOI: 10.1039/D0GC01830A

[70] Carbonell V, Vuorio E, Aro E-M, Kallio P. Enhanced stable production of ethylene in photosynthetic cyanobacterium Synechococcus elongatus PCC 7942. World Journal of Microbiology and Biotechnology. 2019;**35**(5):77. DOI: 10.1007/s11274-019-2652-7

[71] Kobayashi S, Atsumi S, Ikebukuro K, Sode K, Asano R. Light-induced production of isobutanol and 3-methyl-1-butanol by metabolically engineered cyanobacteria. Microbial Cell Factories. 2022;**21**(1):7. DOI: 10.1186/s12934-021-01732-x

[72] Xie H, Kjellström J, Lindblad P. Sustainable production of photosynthetic isobutanol and 3-methyl-1-butanol in the cyanobacterium Synechocystis sp. PCC 6803. Biotechnology for Biofuels and Bioproducts. 2023;**16**(1):134. DOI: 10.1186/s13068-023-02385-1

[73] Liu X, Miao R, Lindberg P, Lindblad P. Modular engineering for

efficient photosynthetic biosynthesis of 1-butanol from CO2 in cyanobacteria. Energy & Environmental Science. 2019;**12**(9):2765–2777. DOI: 10.1039/C9EE01214A

[74] Galetović A, Peña G, Fernández N, Urrutia M, Flores N, Gómez-Silva B, Di Ruggiero J, Shene C, Bustamante M. Cellulose synthase in atacama cyanobacteria and bioethanol production from their exopolysaccharides. Microorganisms. 2023;**11**(11):2668

[75] Böhm J, Kauss K, Michl K, Engelhardt L, Brouwer E-M, Hagemann M. Impact of the carbon flux regulator protein pirC on ethanol production in engineered cyanobacteria. Frontiers in Microbiology. 2023;**14**. DOI: 10.3389/fmicb.2023.1238737

[76] Roussou S, Albergati A, Liang F, Lindblad P. Engineered cyanobacteria with additional overexpression of selected Calvin-Benson-Bassham enzymes show further increased ethanol production. Metabolic Engineering Communications. 2021;**12**:e00161. DOI: 10.1016/j.mec.2021.e00161

[77] Gao E-B, Wu J, Ye P, Qiu H, Chen H, Fang Z. Rewiring carbon flow in Synechocystis PCC 6803 for a high rate of CO2-to-ethanol under an atmospheric environment. Frontiers in Microbiology. 2023;**14**. DOI: 10.3389/fmicb.2023.1211004

[78] Kopka J, Schmidt S, Dethloff F, Pade N, Berendt S, Schottkowski M, Martin N, Dühring U, Kuchmina E, Enke H, Kramer D, Wilde A, Hagemann M, Friedrich A. Systems analysis of ethanol production in the genetically engineered cyanobacterium Synechococcus sp. PCC 7002. Biotechnology for Biofuels. 2017;**10**(1):56. DOI: 10.1186/s13068-017-0741-0

[79] Lin P-C, Zhang F, Pakrasi HB. Enhanced limonene production in a fast-growing cyanobacterium through combinatorial metabolic engineering. Metabolic Engineering Communications. 2021;**12**:e00164. DOI: 10.1016/j.mec.2021.e00164

[80] Shinde S, Singapuri S, Jiang Z, Long B, Wilcox D, Klatt C, Jones JA, Yuan JS, Wang X. Thermodynamics contributes to high limonene productivity in cyanobacteria. Metabolic Engineering Communications. 2022;**14**:e00193. DOI: 10.1016/j.mec.2022.e00193

[81] Vincent M, Blanc-Garin V, Chenebault C, Cirimele M, Farci S, Garcia-Alles LF, Cassier-Chauvat C, Chauvat F. Impact of carbon fixation, distribution and storage on the production of farnesene and limonene in synechocystis PCC 6803 and Synechococcus PCC 7002. International Journal of Molecular Sciences. 2024;**25**(7):3827

[82] Li M, Long B, Dai SY, Golden JW, Wang X, Yuan JS. Altered carbon partitioning enhances CO_2 to terpene conversion in cyanobacteria. BioDesign Research. 2022;**2022**:9897425. DOI: 10.34133/2022/9897425

[83] Lin P-C, Saha R, Zhang F, Pakrasi HB. Metabolic engineering of the pentose phosphate pathway for enhanced limonene production in the cyanobacterium Synechocystis sp PCC 6803. Scientific Reports. 2017;**7**(1):17503. DOI: 10.1038/s41598-017-17831-y

[84] Betterle N, Melis A. Heterologous leader sequences in fusion constructs

enhance expression of geranyl diphosphate synthase and yield of β-phellandrene production in cyanobacteria (Synechocystis). ACS Synthetic Biology. 2018;**7**(3):912–921. DOI: 10.1021/acssynbio.7b00431

[85] Formighieri C, Melis A. A phycocyanin·phellandrene synthase fusion enhances recombinant protein expression and β-phellandrene (monoterpene) hydrocarbons production in Synechocystis (cyanobacteria). Metabolic Engineering. 2015;**32**:116–124. DOI: 10.1016/j.ymben.2015.09.010

[86] Bentley FK, García-Cerdán JG, Chen H-C, Melis A. Paradigm of monoterpene (β-phellandrene) hydrocarbons production via photosynthesis in cyanobacteria. BioEnergy Research. 2013;**6**(3): 917–929. DOI: 10.1007/s12155-013-9325-4

[87] Deng L, Xi H, Wan C, Fu L, Wang Y, Wu C. Is the petrochemical industry an overlooked critical source of environmental microplastics? Journal of Hazardous Materials. 2023;**451**:131199. DOI: 10.1016/j.jhazmat.2023.131199

[88] Tilsted JP, Bauer F, Birkbeck CD, Skovgaard J, Rootzén J. Ending fossil-based growth: Confronting the political economy of petrochemical plastics. One Earth. 2023;**6**(6):607–619. DOI: 10.1016/j.oneear.2023.05.018

[89] Eriksen M, Lebreton LCM, Carson HS, Thiel M, Moore CJ, Borerro JC, Galgani F, Ryan PG, Reisser J. Plastic pollution in the world's oceans: More than 5 trillion plastic pieces weighing over 250,000 Tons afloat at sea. PLOS ONE. 2014;**9**(12): e111913. DOI: 10.1371/journal.pone.0111913

[90] Bergmann M, Collard F, Fabres J, Gabrielsen GW, Provencher JF, Rochman CM, van Sebille E, Tekman MB. Plastic pollution in the Arctic. Nature Reviews Earth & Environment. 2022;**3**(5):323–337. DOI: 10.1038/s43017-022-00279-8

[91] Huang S, Dong Q, Che S, Li R, Tang KHD. Bioplastics and biodegradable plastics: A review of recent advances, feasibility and cleaner production. Science of the Total Environment. 2025;**969**:178911. DOI: 10.1016/j.scitotenv.2025.178911

[92] Mohanasundaram Y, Natarajan V, and Arumugam N. Cyanobacteria as a source of biodegradable plastics. In: Archana T, editor. Cyanobacteria, IntechOpen. Rijeka: IntechOpen; 2023. Ch. 5. DOI: 10.5772/intechopen.110376

[93] Arora Y, Sharma S, Sharma V. Microalgae in bioplastic production: A comprehensive review. Arabian Journal for Science and Engineering. 2023;**48**(6):7225–7241. DOI: 10.1007/s13369-023-07871-0

[94] Altamira-Algarra B, Lage A, Meléndez AL, Arnau M, Gonzalez-Flo E, García J. Bioplastic production by harnessing cyanobacteria-rich microbiomes for long-term synthesis. Science of the Total Environment. 2024;**954**:176136. DOI: 10.1016/j.scitotenv.2024.176136

[95] Nawaz T, Gu L, Hu Z, Fahad S, Saud S, Zhou R. Advancements in synthetic biology for enhancing cyanobacterial capabilities in sustainable plastic production: A green horizon perspective. Fuels. 2024;**5**(3):394–438

[96] Rueda E, Gonzalez-Flo E, Mondal S, Forchhammer K, Arias DM,

Ludwig K, Drosg B, Fritz I, Gonzalez-Esquer CR, Pacheco S, García J. Challenges, progress, and future perspectives for cyanobacterial polyhydroxyalkanoate production. Reviews in Environmental Science and Bio/Technology. 2024;**23**(2): 321–350. DOI: 10.1007/s11157-024-09689-0

[97] Aluru R, RinkuPolachirakkal V. A review of PHB production by cyanobacteria and its applications. Journal of Biochemical Technology. 2022;**13**(4):50–53. DOI: 10.51847/ty7wnH5iA3

[98] Altamira-Algarra B, Rueda E, Lage A, León DS, Martínez-Blanch JF, Nogales J, García J, Gonzalez-Flo E. New strategy for bioplastic and exopolysaccharides production: Enrichment of field microbiomes with cyanobacteria. New Biotechnology. 2023;**78**:141–149. DOI: 10.1016/j.nbt.2023.10.008

[99] Fink P, Menzel C, Kwon J-H, Forchhammer K. A novel recombinant PHB production platform in filamentous cyanobacteria avoiding nitrogen starvation while preserving cell viability. Microbial Cell Factories. 2025;**24**(1):43. DOI: 10.1186/s12934-025-02650-y

[100] Gradíssimo DG, Xavier LP, Santos AV.Cyanobacterial Polyhydroxyalkanoates: A sustainable alternative in circular economy. Molecules. 2020;**25**(18). DOI: 10.3390/molecules25184331

[101] Khetkorn W, Incharoensakdi A, Lindblad P, Jantaro S. Enhancement of poly-3-hydroxybutyrate production in Synechocystis sp. PCC 6803 by overexpression of its native biosynthetic genes. Bioresource Technology. 2016;**214**:761–768. DOI: 10.1016/j.biortech.2016.05.014

[102] Kamravamanesh D, Pflügl S, Nischkauer W, Limbeck A, Lackner M, Herwig C. Photosynthetic poly-β-hydroxybutyrate accumulation in unicellular cyanobacterium Synechocystis sp PCC 6714. AMB Express. 2017;**7**(1):143. DOI: 10.1186/s13568-017-0443-9

[103] Koch M, Bruckmoser J, Scholl J, Hauf W, Rieger B, Forchhammer K. Maximizing PHB content in Synechocystis sp. PCC 6803: A new metabolic engineering strategy based on the regulator PirC. Microbial Cell Factories. 2020;**19**(1):231. DOI: 10.1186/s12934-020-01491-1

[104] Djebaili R, Mignini A, Vaccarelli I, Pellegrini M, Spera DM, Gallo MD, D'Alessandro AM. Polyhydroxybutyrate-producing cyanobacteria from lampenflora: The case study of the "Stiffe". Caves in Italy, Frontiers in Microbiology. 2022;**13**. DOI: 10.3389/fmicb.2022.933398

[105] Samantaray S, Mallick N. Production and characterization of poly-β-hydroxybutyrate (PHB) polymer from Aulosira fertilissima. Journal of Applied Phycology. 2012;**24**(4):803–814. DOI: 10.1007/s10811-011-9699-7

[106] Hirai K, Nojo M, Sato Y, Tsuzuki M, Sato N. Contribution of protein synthesis depression to poly-β-hydroxybutyrate accumulation in Synechocystis sp. PCC 6803 under nutrient-starved conditions. Scientific Reports. 2019;**9**(1):19944. DOI: 10.1038/s41598-019-56520-w

[107] Mittermair S, Richter J, Doppler P, Trenzinger K, Nicoletti C,

Forsich C, Spadiut O, Herwig C, Lackner M. Impact of exoD gene knockout on the polyhydroxybutyrate overaccumulating mutant Mt_a24. International Journal of Biobased Plastics. 2021;**3**(1):1–18. DOI: 10.1080/24759651.2020.1863020

[108] Price S, Kuzhiumparambil U, Pernice M, Herdean A, Ralph P. Enhancement of cyanobacterial PHB production using random chemical mutagenesis with detection through FACS. Bioprocess and Biosystems Engineering. 2023;**46**(2):297–306. DOI: 10.1007/s00449-022-02834-5

[109] Bai X, Zhang G, Xu L, Li K, Zhang M, Huang Y. Rewiring photosynthetic carbon flow: Engineered cyanobacterial factories for sustainable carbohydrate production and carbon-negative biomanufacturing. Bioresource Technology. 2025;**434**:132850. DOI: 10.1016/j.biortech.2025.132850

[110] Li X, Liang G, Liang Y, Kong F, Chi Z. Engineered cyanobacteria for improving poly-β-hydroxybutyrate production under NaHCO3 treatment. Environmental Research. 2025;**280**:121894. DOI: 10.1016/j.envres.2025.121894

[111] Koch M, Berendzen KW, Forchhammer K. On the role and production of Polyhydroxybutyrate (PHB) in the cyanobacterium synechocystis sp.PCC 6803. Life. 2020;**10**(4):47

[112] Rueda E, Altamira-Algarra B, García J. Process optimization of the polyhydroxybutyrate production in the cyanobacteria Synechocystis sp. and Synechococcus sp. Bioresource Technology. 2022;**356**:127330. DOI: 10.1016/j.biortech.2022.127330

[113] Meixner K, Daffert C, Bauer L, Drosg B, Fritz I. PHB producing cyanobacteria found in the neighborhood—Their isolation, purification and performance testing. Bioengineering. 2022;**9**(4):178

[114] Ansari S, Fatma T. Cyanobacterial Polyhydroxybutyrate (PHB): Screening, optimization and characterization. PLOS ONE. 2016;**11**(6):e0158168. DOI: 10.1371/journal.pone.0158168

[115] Simonazzi M, Pezzolesi L, Galletti P, Gualandi C, Pistocchi R, De Marco N, Paganelli Z, Samorì C. Production of polyhydroxybutyrate by the cyanobacterium cf. Anabaena Sp, International Journal of Biological Macromolecules. 2021;**191**:92–99. DOI: 10.1016/j.ijbiomac.2021.09.054

[116] Rueda E, García-Galán MJ, Díez-Montero R, Vila J, Grifoll M, García J. Polyhydroxybutyrate and glycogen production in photobioreactors inoculated with wastewater borne cyanobacteria monocultures. Bioresource Technology. 2020;**295**:122233. DOI: 10.1016/j.biortech.2019.122233

[117] Carpine R, Olivieri G, Hellingwerf KJ, Pollio A, Pinto G, Marzocchella A. Photoautotrophic production of poly-.-hydroxybutyrate (phb) from cyanobacteria: Nitrate effects and screening of strains. Environmental Engineering and Management Journal. 2019;**18**(6):10

[118] Omar HH, Aly MM, Al-Malki WJ, Balkhair KS. Production and enhancement of poly-β-hydroxybutyrate in cyanobacteria. Main Group Chemistry. 2016;**15**(2):153–161. DOI: 10.3233/mgc-150194

[119] Nguyen TLA, Tran HTT, Quach TMT, Dao YH, Duong HC, Doan OT, Duong TT, Tran LTT. Biopolymer polyhydroxyalkanoate production from Arthrospira platensis NLHT3 cyanobacterium isolated in Vietnam. Environmental Technology & Innovation. 2024;**36**:103841. DOI: 10.1016/j.eti.2024.103841

[120] Dhankhar N, Kumar J, Impact of increasing pesticides and fertilizers on human health: A review. In: Materials Today: Proceedings; (2023); DOI: 10.10 16/j.matpr.2023.03.766

[121] Ahmad MF, Ahmad FA, Alsayegh AA, Zeyaullah M, AlShahrani AM, Muzammil K, Saati AA, Wahab S, Elbendary EY, Kambal N, Abdelrahman MH, Hussain S. Pesticides impacts on human health and the environment with their mechanisms of action and possible countermeasures. Heliyon. 2024;**10**(7):e29128. DOI: 10.1016/j.heliyon.2024.e29128

[122] Gamage A, Gangahagedara R, Gamage J, Jayasinghe N, Kodikara N, Suraweera P, Merah O. Role of organic farming for achieving sustainability in agriculture. Farming System. 2023;**1**(1):100005. DOI: 10.1016/j.farsys.2023.100005

[123] Agarwal P, Soni R, Kaur P, Madan A, Mishra R, Pandey J, Singh S, Singh G. Cyanobacteria as a promising alternative for sustainable environment: synthesis of biofuel and biodegradable plastics. Frontiers in Microbiology. 2022;**13**:939347. DOI: 10.3389/fmicb.2022.939347

[124] Bibi S, Saadaoui I, Bibi A, Al-Ghouti M, Abu-Dieyeh MH. Applications, advancements, and challenges of cyanobacteria-based biofertilizers for sustainable agro and ecosystems in arid climates. Bioresource Technology Reports. 2024;**25**:101789. DOI: 10.1016/j.biteb.2024.101789

[125] Robati SMS. Cyanobacteria: A promising future for sustainable agriculture. In: Severo IA, Martínez-Burgos WJ, Ordonez J, editors. Insights into Algae - Fundamentals, Culture Techniques and Biotechnological Uses of Microalgae and Cyanobacteria. Rijeka: IntechOpen. 2024:77–96. DOI: 10.5772/intechopen.1005021

[126] Nawaz T, Fahad S, Gu L, Xu L, Zhou R. Harnessing nitrogen-fixing cyanobacteria for sustainable agriculture: Opportunities, challenges, and implications for food security. Nitrogen. 2025;**6**(1):16

[127] Pandey P, Pandey D, Gupta A, Gupta R, Tiwari S, Singh SP. Cyanobacterial green chemistry: A blue-green approach for a sustainable environment, energy, and chemical production. RSC Sustainability. 2025;**3**(2):661–675. DOI: 10.1039/D4SU00448E

[128] Hata S, Kishida S, Minesono R, Tamai T. Dried Nostoc commune exhibits nitrogen-fixing activity using glucose under dark conditions after rehydration. Plant Signaling & Behavior. 2022;**17**(1):2059251. DOI: 10.1080/15592324.2022.2059251

[129] Jiménez-Ríos L, Torrado A, González-Pimentel JL, Iniesta-Pallarés M, Molina-Heredia FP, Mariscal V, Álvarez C. Emerging nitrogen-fixing cyanobacteria for sustainable cotton cultivation. Science of the Total Environment. 2024;**924**:171533. DOI: 10.1016/j.scitotenv.2024.171533

[130] Chatterjee P, Schafran P, Li F-W, Meeks JC. Nostoc talks back: Temporal

patterns of differential gene expression during establishment of anthoceros-nostoc symbiosis. Molecular Plant-Microbe Interactions®. 2022;**35**(10):917–932. DOI: 10.1094/mpmi-05-22-0101-r

[131] Lucato V, Ihadjadene Y, Sut S, Dall'Acqua S, Sforza E, Krujatz F. Optimization of Nostoc sp. biomass and protein composition in high-density cultivator based on DoE. Algal Research. 2025;**89**:104088. DOI: 10.1016/j.algal.2025.104088

[132] Brocke HJ, Piltz B, Herz N, Abed RMM, Palinska KA, John U, Haan JD, de Beer D, Nugues MM. Nitrogen fixation and diversity of benthic cyanobacterial mats on coral reefs in Curaçao. Coral Reefs. 2018;**37**(3):861–874. DOI: 10.1007/s00338-018-1713-y

[133] Tuo S-H, Chen Y-LL, Chen H-Y, Chen T-Y. Free-living heterocystous cyanobacteria in the tropical marginal seas of the western North Pacific. Journal of Plankton Research. 2017;**39**(3):404–422. DOI: 10.1093/plankt/fbx023

[134] Berrendero E, Valiente EF, Perona E, Gómez CL, Loza V, Muñoz-Martín MÁ, Mateo P. Nitrogen fixation in a non-heterocystous cyanobacterial mat from a mountain river. Scientific Reports. 2016;**6**(1):30920. DOI: 10.1038/srep30920

[135] Li H, Miller T, Lu J, Goel R. Nitrogen fixation contribution to nitrogen cycling during cyanobacterial blooms in Utah Lake. Chemosphere. 2022;**302**:134784. DOI: 10.1016/j.chemosphere.2022.134784

[136] Yoon K-S, Nguyen NT, Tran KT, Tsuji K, Ogo S. Nitrogen fixation genes and nitrogenase activity of the non-heterocystous cyanobacterium *thermoleptolyngbya* sp. O-77. Microbes and Environments. 2017;**32**(4):324–329. DOI: 10.1264/jsme2.ME17015

[137] Sadvakasova AK, Kossalbayev BD, Token AI, Bauenova MO, Wang J, Zayadan BK, Balouch H, Alwasel S, Leong YK, Chang J-S, Allakhverdiev SI. Influence of Mo and Fe on photosynthetic and nitrogenase activities of nitrogen-fixing cyanobacteria under nitrogen starvation. Cells. 2022;**11**(5):904

[138] Stal LJ. The effect of oxygen concentration and temperature on nitrogenase activity in the heterocystous cyanobacterium Fischerella sp. Scientific Reports. 2017;7(1):5402. DOI: 10.1038/s41598-017-05715-0

[139] Singh K.Determination of nitrogenase activity in Stigonematales. Journal of Natural Resource and Development. 2016;**11**(1):1–5

[140] Wu L, Dong J, Song J, Zhu Y, Che S, Qin X, Xu Y, Tian S, Wang D, Tian P, Wei X, Yang X, Yang M, Wu Z. The nitrogen fixation characteristics of terrestrial nitrogen-fixing cyanobacteria and their role in promoting rice growth. Agronomy. 2025;**15**(1):62

[141] Yadav S, Rai S, Rai R, Shankar A, Singh S, Rai LC. Cyanobacteria: Role in agriculture, environmental sustainability, biotechnological potential and agroecological impact. In: Singh DP, Singh HB, Prabha R, editors. Plant-Microbe Interactions in Agro-Ecological Perspectives: Volume 2: Microbial Interactions and Agro-Ecological Impacts. Singapore, Singapore: Springer; 2017. p. 257–277. DOI: 10.1007/978-981-10-6593-4_10

[142] Pathak J, Rajneesh PKM, Singh SP, Häder D-P, Sinha RP. Cyanobacterial farming for environment friendly sustainable agriculture practices: innovations and perspectives. Frontiers in Environmental Science. 2018;**6**. DOI: 10.3389/fenvs.2018.00007

[143] Gupta C, Rabani MS, Gupta MK, Habib A, Pathak A, Tripathi S, Singh R. Cyanobacteria as sustainable microbiome for agricultural industries. In: Dar GH, Bhat RA, Mehmood MA, Hakeem KR, editors. Microbiota and Biofertilizers, Vol 2: Ecofriendly Tools for Reclamation of Degraded Soil Environs. Cham: Springer International Publishing; 2021. p. 267–291. DOI: 10.1007/978-3-030-6101 0-4_13

[144] Stal LJ. The effect of oxygen concentration and temperature on nitrogenase activity in the heterocystous cyanobacterium Fischerella sp. Scientific Reports. 2017;**7**(1):5402. DOI: 10.1038/s4 1598-017-05715-0

[145] Bandyopadhyay A, Elvitigala T, Liberton M, Pakrasi HB. Variations in the rhythms of respiration and nitrogen fixation in members of the unicellular diazotrophic cyanobacterial genus Cyanothece. Plant Physiology. 2013;**161** (3):1334–1346. DOI: 10.1104/ pp.112.208231

[146] Stal LJ. Nitrogen fixation in cyanobacteria. In: Encyclopedia of Life Sciences. Hoboken, NJ, USA: Wiley; 2015. p. 1–9. DOI: 10.1002/9780470015 902.a0021159.pub2

[147] Liu D, Liberton M, Yu J, Pakrasi HB, Bhattacharyya-Pakrasi M. Engineering nitrogen fixation activity in an oxygenic phototroph. mBio. 2018;**9** (3). DOI: 10.1128/mbio.01029-18

[148] Bennett EM, Murray JW, Isalan M. Engineering nitrogenases for synthetic nitrogen fixation: From pathway engineering to directed evolution. BioDesign Research. 2023;**5**:0005. DOI: 10.34133/bdr.0005

[149] Pérez Gallego R, Bale NJ, Damste JSS, Villanueva L. Developing a genetic approach to target cyanobacterial producers of heterocyte glycolipids in the environment. Frontiers in Microbiology. 2023;**14**. DOI: 10.3389/fmicb.2023.1257040

[150] Werner T, Nehnevajova E, Köllmer I, Novák O, Strnad M, Krämer U, Schmülling T. Root-Specific reduction of cytokinin causes enhanced root growth, drought tolerance. and Leaf Mineral Enrichment in Arabidopsis and Tobacco the Plant Cell. 2010;**22** (12):3905–3920. DOI: 10.1105/ tpc.109.072694

[151] Finkelstein R. Abscisic acid synthesis and response. The Arabidopsis Book. 2013;**1**(11):e0166

[152] Gupta R, Chakrabarty SK. Gibberellic acid in plant. Plant Signaling & Behavior. 2013;**8**(9):e25504. DOI: 10. 4161/psb.25504

[153] Li S-W, Zeng X-Y, Leng Y, Feng L, Kang X-H. Indole-3-butyric acid mediates antioxidative defense systems to promote adventitious rooting in mung bean seedlings under cadmium and drought stresses. Ecotoxicology and Environmental Safety. 2018;**161**:332–341. DOI: 10.1016/j.ecoenv.2018.06.003

[154] Carpine R, Sieber S. Antibacterial and antiviral metabolites from cyanobacteria: Their application and their impact on human health. Current Research in Biotechnology. 2021;**3**:65–81. DOI: 10.1016/j.crbiot.2021.03.001

[155] Kar J, Ramrao DP, Zomuansangi R, Lalbiaktluangi C, Singh SM, Joshi NC, Kumar A, Kaushalendra SM, Yadav MK, Singh PK. Revisiting the role of cyanobacteria-derived metabolites as antimicrobial agent: A 21st century perspective. Frontiers in Microbiology. 2022;**13**. DOI: 10.3389/fmicb.2022.1034471

[156] Kim J, Kim JD. Inhibitory effect of algal extracts on mycelial growth of the tomato-wilt pathogen Fusarium oxysporum f. sp. lycopersici. Mycobiology. 2008;**36**(4):242–248. DOI: 10.4489/myco.2008.36.4.242

[157] Ghareeb RY, Abdelsalam NR, Maghraby DME, Ghozlan MH, EL-Argawy E. Abou-Shanab RAI, Oscillatoria sp. as a potent anti-phytopathogenic agent and plant immune stimulator against root-knot nematode of soybean cv Giza 111. Frontiers in Plant Science. 2022;**13**. DOI: 10.3389/fpls.2022.870518

[158] Massa A, Santos É, Martins D, Azevedo J, Reimão M, Almeida A, Azevedo R, Pinto E, Vasconcelos V, Campos A, Freitas M. Toxic and non-toxic cyanobacterial biomass as a resource for sustainable agriculture: A lettuce cultivation experiment. Environmental Research. 2024;**262**:119942. DOI: 10.1016/j.envres.2024.119942

[159] Singh L, Nayak A, Behera M, Sethi G, Alheswairini SS, Barasarathi J, Sayyed R, Behera DK, Adarsh V. Antifungal potential of cyanobacterium desertifilum dzianense against collar rot pathogen in Chickpea. Scientific Reports. 2025;**15**(1):23740. DOI: 10.1038/s41598-025-08395-3

[160] Li H, Zhao Q, Huang H. Current states and challenges of salt-affected soil remediation by cyanobacteria. Science of the Total Environment. 2019;**669**:258–272. DOI: 10.1016/j.scitotenv.2019.03.104

[161] Duan H, Liu W, Zhou L, Han B, Huo S, El-Sheekh M, Dong H, Li X, Xu T, Elshobary M. Improving saline-alkali soil and promoting wheat growth by co-applying potassium-solubilizing bacteria and cyanobacteria produced from brewery wastewater. Frontiers in Environmental Science. 2023;**11**. DOI: 10.3389/fenvs.2023.1170734

[162] Ramakrishnan B, Maddela NR, Venkateswarlu K, Megharaj M. Potential of microalgae and cyanobacteria to improve soil health and agricultural productivity: A critical view. Environmental Science: Advances. 2023;**2**(4):586–611. DOI: 10.1039/D2VA00158F

[163] Gufwan LA, Peng L, Gufwan NM, Lan S, Wu L. Enhancing soil health through biocrusts: A microbial ecosystem approach for degradation control and restoration. Microbial Ecology. 2025;**88**(1):8. DOI: 10.1007/s00248-025-02504-5

[164] Wood R. Acute animal and human poisonings from cyanotoxin exposure — A review of the literature. Environment International. 2016;**91**:276–282. DOI: 10.1016/j.envint.2016.02.026

[165] Villalobos T, Suárez-Isla B, Garcia C. health and environmental impacts of cyanobacteria and cyanotoxins from freshwater to seawater. Toxins. 2025;**17**(3):126

[166] Duong TT, Nguyen TTL, Dinh THV, Hoang TQ, Vu TN, Doan TO, Dang TMA, Le TPQ,

Tran DT, Le VN, Nguyen QT, Le PT, Nguyen TK, Pham TD, Bui HM. Auxin production of the filamentous cyanobacterial Planktothricoides strain isolated from a polluted river in Vietnam. Chemosphere. 2021;**284**:131242. DOI: 10.1016/j.chemosphere.2021.131242

[167] Mazhar S, Cohen JD, Hasnain S. Auxin producing non-heterocystous Cyanobacteria and their impact on the growth and endogenous auxin homeostasis of wheat. Journal of Basic Microbiology. 2013;**53**(12):996–1003. DOI: 10.1002/jobm.201100563

[168] Žižková E, Kubeš M, Dobrev PI, Přibyl P, Šimura J, Zahajská L, Záveská Drábková L, Novák O, Motyka V. Control of cytokinin and auxin homeostasis in cyanobacteria and algae. Annals of Botany. 2016;**119**(1):151–166. DOI: 10.1093/aob/mcw194

[169] Toribio AJ, Suárez-Estrella F, Jurado MM, López MJ, López-González JA, Moreno J. Prospection of cyanobacteria producing bioactive substances and their application as potential phytostimulating agents. Biotechnology Reports. 2020;**26**: e00449. DOI: 10.1016/j.btre.2020.e00449

[170] Maršálek B, Zahradníčková H, Hronková M. Extracellular Abscisic Acid Produced by Cyanobacteria under Salt Stress. Journal of Plant Physiology. 1992;**139**(4):506–508. DOI: 10.1016/S0176-1617(11)80503-1

[171] Touliabah HE, El-Sheekh MM, Ismail MM, El-Kassas H. A review of microalgae- and cyanobacteria-based biodegradation of organic pollutants. Molecules. 2022;**27**(3). DOI: 10.3390/molecules27031141

[172] Behl K, Devi A, Yadav Y, and Jaiswal P. Chapter 14 - Cyanobacteria: A precious bioresource for bioremediation. In: Mishra AK, and Singh SS, editors. Cyanobacteria. London, UK: Academic Press; 2024. p. 341–382. DOI: 10.1016/B978-0-443-13231-5.00013-1

[173] Dudeja C, Masroor S, Mishra V, Kumar K, Sansar S, Yadav P, Chaturvedi N, Singh R, Kumar A. Cyanobacteria-based bioremediation of environmental contaminants: Advances and computational insights. Discover Agriculture. 2025;**3**(1):42. DOI: 10.1007/s44279-025-00193-9

[174] Acharya C, Chandwadkar P, Apte SK. Interaction of uranium with a filamentous, heterocystous, nitrogen-fixing cyanobacterium, Anabaena torulosa. Bioresource Technology. 2012;**116**:290–294. DOI: 10.1016/j.biortech.2012.03.068

[175] Singh JS, Singh DP, and Dixit S. Cynobacteria: An agent of heavy metal removal. In: Bioremediation of Pollutants (New Delhi: I K International Publisher Co). 2012. 223–243

[176] Hamouda RAEF, Sorour NM, Yeheia DS. Biodegradation of crude oil by Anabaena oryzae, Chlorella kessleri and its consortium under mixotrophic conditions. International Biodeterioration & Biodegradation. 2016;**112**:128–134. DOI: 10.1016/j.ibiod.2016.05.001

[177] Kumar D, Pandey LK, Gaur JP. Metal sorption by algal biomass: From batch to continuous system. Algal Research. 2016;**18**:95–109. DOI: 10.1016/j.algal.2016.05.026

[178] Shen L, Xia J-L, He H, Nie Z-Y. Comparative study on biosorption of Pb(II) and Cr(VI) by Synechococcus sp.

Transactions of Nonferrous Metals Society of China. 2008;**18**(6):1336–1342. DOI: 10.1016/S1003-6326(09)60006-6

[179] Azizi SN, Colagar AH, Hafeziyan SM. Removal of Cd(II) from aquatic system using oscillatoria sp. biosorbent. The Scientific World Journal. 2012;**2012**(1):347053. DOI: 10.1100/2012/347053

[180] Sun X, Huang H, Zhu Y, Du Y, Yao L, Jiang X, Gao P. Adsorption of Pb2 + and Cd2+ onto Spirulina platensis harvested by polyacrylamide in single and binary solution systems. Colloids and Surfaces A: Physicochemical and Engineering Aspects. 2019;**583**:123926. DOI: 10.1016/j.colsurfa.2019.123926

[181] Zeng G, He Y, Liang D, Wang F, Luo Y, Yang H, Wang Q, Wang J, Gao P, Wen X, Yu C, Sun D. Adsorption of heavy metal ions copper, cadmium and nickel by microcystis aeruginosa. International Journal of Environmental Research and Public Health. 2022;**19**(21):13867

[182] Hossain MS, Okino T. Cyanoremediation of heavy metals (As(v), Cd(ii), Cr(vi), Pb(ii)) by live cyanobacteria (Anabaena variabilis, and Synechocystis sp.): An eco-sustainable technology. RSC Advances. 2024;**14**(15):10452–10463. DOI: 10.1039/D4RA00409D

[183] Pagli C, Chamizo S, Migliore G, Rugnini L, De Giudici G, Braglia R, Canini A, Cantón Y. Isolation of biocrust cyanobacteria and evaluation of Cu, Pb, and Zn immobilisation potential for soil restoration and sustainable agriculture. Science of the Total Environment. 2024;**946**:174020. DOI: 10.1016/j.scitotenv.2024.174020

[184] Abdelkarim MS, Ali MHH, Kassem DA. Ecofriendly remediation of cadmium, lead, and zinc using dead cells of Microcystis aeruginosa. Scientific Reports. 2025;**15**(1):3677. DOI: 10.1038/s41598-025-86884-1

[185] Kalita N, Baruah PP. Cyanobacteria as a potent platform for heavy metals biosorption: Uptake, responses and removal mechanisms. Journal of Hazardous Materials Advances. 2023;**11**:100349. DOI: 10.1016/j.hazadv.2023.100349

[186] Bestawy EE. Efficiency of immobilized cyanobacteria in heavy metals removal from industrial effluents. Desalination and Water Treatment. 2019;**159**:66–78. DOI: 10.5004/dwt.2019.23808

[187] Bloch K, and Ghosh S. Chapter 23 - Cyanobacteria mediated toxic metal removal as complementary and alternative wastewater treatment strategy. In: Kumar V, and Kumar M, editors. Integrated Environmental Technologies for Wastewater Treatment and Sustainable Development. Amsterdam, Netherlands: Elsevier; 2022. p. 533–548. DOI: 10.1016/B978-0-323-91180-1.00002-8

[188] Goswami RK, Agrawal K, Shah MP, Verma P. Bioremediation of heavy metals from wastewater: A current perspective on microalgae-based future. Letters in Applied Microbiology. 2022;**75**(4):701–717. DOI: 10.1111/lam.13564

[189] Mahlangu D, Mphahlele K, De Paola F, Mthombeni NH. Microalgae-Mediated biosorption for effective heavy metals removal from wastewater: A review. Water. 2024;**16**(5):718

[190] Rahman MA, Soumya KK, Tripathi A, Sundaram S, Singh S,

Gupta A. Evaluation and sensitivity of cyanobacteria, Nostoc muscorum and Synechococcus PCC 7942 for heavy metals stress – A step toward biosensor. Toxicological & Environmental Chemistry. 2011;**93**(10):1982–1990. DOI: 10.1080/02772248.2011.606110

[191] Pandi M, Shashirekha V, Swamy M. Bioabsorption of chromium from retan chrome liquor by cyanobacteria. Microbiological Research. 2009;**164**(4):420–428. DOI: 10.1016/j.micres.2007.02.009

[192] Sen G, Sen S, Thakurta SG, Chakrabarty J, Dutta S. Bioremediation of Cr(VI) Using live cyanobacteria: Experimentation and kinetic modeling. Journal of Environmental Engineering. 2018;**144**(9):04018089. DOI: 10.1061/(ASCE)EE.1943-7870.0001425

[193] Ibrahim WM, Karam MA, El-Shahat RM, Adway AA. Biodegradation and utilization of organophosphorus pesticide malathion by cyanobacteria. BioMed Research International. 2014;**2014**(1):392682. DOI: 10.1155/2014/392682

[194] Tiwari B, Chakraborty S, Srivastava AK, Mishra AK. Biodegradation and rapid removal of methyl parathion by the paddy field cyanobacterium Fischerella sp. Algal Research. 2017;**25**:285–296. DOI: 10.1016/j.algal.2017.05.024

[195] Sarasa-Buisán C, Guío J, Castro C, Bes MT, Fillat MF, Peleato ML, Sevilla E. Contributions on lindane degradation by microcystis aeruginosa PCC 7806. Water. 2022;**14**(8):1219

[196] Abou Elatta AEA, Zawawy HAHE, Afify AH, Hauka FIA. Degradation of chlorpyrifos by the cyanobacteria strains in rice fields. Journal of Agricultural Chemistry and Biotechnology. 2023;**14**(5):43–49. DOI: 10.21608/jacb.2023.202369.1050

[197] González-Barreiro O, Rioboo C, Herrero C, Cid A. Removal of triazine herbicides from freshwater systems using photosynthetic microorganisms. Environmental Pollution. 2006;**144**(1):266–271. DOI: 10.1016/j.envpol.2005.12.014

[198] Kumar JIN, Amb MK, Bora A. Chronic response of Anabaena fertilissima Rao, C.B. on growth, metabolites and enzymatic activities by chlorophenoxy herbicide. Pesticide Biochemistry and Physiology. 2010;**98**(2):168–174. DOI: 10.1016/j.pestbp.2010.06.002

[199] Forlani G, Prearo V, Wieczorek D, Kafarski P, Lipok J. Phosphonate degradation by Spirulina strains: Cyanobacterial biofilters for the removal of anticorrosive polyphosphonates from wastewater. Enzyme and Microbial Technology. 2011;**48**(3):299–305. DOI: 10.1016/j.enzmictec.2010.12.005

[200] Campos MMC, Faria VHF, Teodoro TS, Barbosa FAR, Magalhães SMS. Evaluation of the capacity of the cyanobacterium Microcystis novacekii to remove atrazine from a culture medium. Journal of Environmental Science and Health, Part B. 2013;**48**(2):101–107. DOI: 10.1080/03601234.2013.726891

[201] Agrawal C, Sen S, Yadav S, Rai S, Rai LC. A novel aldo-keto reductase (AKR17A1) of Anabaena sp. PCC 7120 degrades the rice field herbicide butachlor and confers tolerance to abiotic stresses in E. coli. PLOS ONE.

2015;**10**(9):e0137744. DOI: 10.1371/journal.pone.0137744

[202] Sachu M, Kynshi BL, Syiem MB. Cyanobacterial degradation of herbicide 2,4-dichlorophenoxyacetic acid (2,4-D): Its response to the oxidative stress induced by the primary degradation product 2,4-dichlorophenol (2,4-DCP). Comparative Biochemistry and Physiology Part C: Toxicology & Pharmacology. 2023;**273**:109739. DOI: 10.1016/j.cbpc.2023.109739

[203] Hamai-Amara H, Saadaoui I, Cherif M, Da'ana DA, Soubra L, Al-Ghouti MA. Evidencing nickel biosorption capacity of cyanobacteria Chroococcidiopsis sp.: Potential metallo-protective agents. BMC Chemistry. 2025;**19**(1):59. DOI: 10.1186/s13065-025-01393-6

[204] Zinicovscaia I, Yushin N, Gundorina S, Demčák Š, Frontasyeva M, Kamanina I. Biosorption of nickel from model solutions and electroplating industrial effluent using cyanobacterium Arthrospira platensis. Desalination and Water Treatment. 2018;**120**:158–165. DOI: 10.5004/dwt.2018.22691

[205] Rakic I, Kevrešan Ž, Kovač R, Kravić S, Svircev Z, Djurovic A, Stojanovic Z. Bioaccumulation and biosorption study of heavy metals removal by Cyanobacteria Nostoc sp. Chemical Industry and Chemical Engineering Quarterly. 2023;**29**:2–2. DOI: 10.2298/CICEQ220511002R

[206] Safari M, Ahmady-Asbchin S. Biosorption of zinc from aqueous solution by cyanobacterium Fischerella ambigua ISC67: Optimization, kinetic, isotherm and thermodynamic studies. Water Science and Technology: a Journal of the International Association on Water Pollution Research. 2018;**78**(7):1525–1534. DOI: 10.2166/wst.2018.437

[207] Cepoi L, Zinicovscaia I, Valuta A, Codreanu L, Rudi L, Chiriac T, Yushin N, Grozdov D, Peshkova A. Peculiarities of the edaphic cyanobacterium Nostoc linckia culture response and heavy metal accumulation from copper-containing multimetal systems. Toxics. 2022;**10**(3):113

[208] Cheraghpour J, Etemadifar Z, Afsharzadeh S, Bahador N. Assessment of bioremediation potential of Microcystis aeruginosa for removal of cadmium and lead ions from aqueous matrices. Iranian Journal of Fisheries Sciences. 2020;**19**(4):1994–2009. DOI: 10.22092/ijfs.2018.119887

[209] Rzymski P, Niedzielski P, Karczewski J, Poniedziałek B. Biosorption of toxic metals using freely suspended Microcystis aeruginosa biomass. Open Chemistry. 2014;**12**(12):1232–1238. DOI: 10.2478/s11532-014-0576-5

[210] Ciani M, Lepore GO, Puri A, Facchetti G, Adessi A. Exploring metal interactions with released polysaccharides from cyanothece sp. CE4: A chemical and spectroscopic study on biosorption mechanism. Polymers. 2025;**17**(3):371

[211] Shen L, Chen R, Wang J, Fan L, Cui L, Zhang Y, Cheng J, Wu X, Li J, Zeng W. Biosorption behavior and mechanism of cadmium from aqueous solutions by Synechocystis sp. PCC6803. RSC Advances. 2021;**11**(30):18637–18650. DOI: 10.1039/D1RA02366G

[212] El-Hameed MMA, Abuarab ME, Al-Ansari N, Mottaleb SA, Bakeer GA,

Gyasi-Agyei Y, Mokhtar A. Phycoremediation of contaminated water by cadmium (Cd) using two cyanobacterial strains (Trichormus variabilis and Nostoc muscorum). Environmental Sciences Europe. 2021;**33**(1):135. DOI: 10.1186/s12302-021-00573-0

[213] Alwaleed EA, Latef AAA, El-Sheekh M. Biosorption efficacy of living and non-living algal cells of Microcystis aeruginosa to toxic metals. Notulae Botanicae Horti Agrobotanici Cluj-Napoca. 2021;**49**(1):12149. DOI: 10.15835/nbha49112149

[214] Paper M, Koch M, Jung P, Lakatos M, Nilges T, Brück TB. Rare earths stick to rare cyanobacteria: Future potential for bioremediation and recovery of rare earth elements. Frontiers in Bioengineering and Biotechnology. 2023;**11**. DOI: 10.3389/fbioe.2023.1130939

[215] Smječanin N, Nuhanović M, Sulejmanović J, Mašić E, Sher F. Highly effective sustainable membrane based cyanobacteria for uranium uptake from aqueous environment. Chemosphere. 2023;**313**:137488. DOI: 10.1016/j.chemosphere.2022.137488

[216] Yu R, Chai H, Yu Z, Wu X, Liu Y, Shen L, Li J, Ye J, Liu D, Ma T, Gao F, Zeng W. Behavior and mechanism of cesium biosorption from aqueous solution by living synechococcus PCC7002. Microorganisms. 2020;**8**(4):491

[217] Zinicovscaia I, Safonov A, Zelenina D, Ershova Y, Boldyrev K. Evaluation of biosorption and bioaccumulation capacity of cyanobacteria Arthrospira (spirulina) platensis for radionuclides. Algal Research. 2020;**51**:102075. DOI: 10.1016/j.algal.2020.102075

Chapter 3

Structural and Functional Roles of the Inorganic Cofactors of Photosystem II

Leslie Castillo, Nidhi Patel, Nicole Seliga, Gustavo Chavez and Colin Gates

Abstract

Photosystem II (PSII) is the sole water oxidase in nature. This large, multisubunit enzyme complex relies on several critical inorganic cofactors for its operation, namely manganese, calcium, chloride, bicarbonate, and iron. In this review, we address the state of the field with regard to the roles of these cofactors in the operation and functionalization of PSII. We specifically address recent developments in our understanding of the water-oxidizing complex (WOC), the catalytic site of water oxidation, and the processes regulating the oxidation states of the catalytic manganese and the role of redox-inert calcium; the function of chloride near the WOC; and the role of the nonheme iron and its necessary bicarbonate ligand in regulating electron transfer between the reaction center and the terminal electron acceptor Q_B.

Keywords: Photosystem II, electron transport, water oxidation, chloride, bicarbonate

1. Introduction

Oxygenic photosynthesis is the fundamental biological process that sustains nearly all life on Earth by converting sunlight into chemical energy. This innovation transformed Earth's atmosphere and enabled the evolution of aerobic life [1]. Oxygenic photosynthesis is carried out by diverse organisms, including cyanobacteria, algae, and plants, whose ability to split water and power electron transport drives global oxygen and biomass production. The capacity of phototrophs to oxidize water and drive linear electron transport originates from Photosystem II (PSII), a large multisubunit protein complex embedded in the thylakoid membrane [2]. PSII catalyzes the light-driven oxidation of water at the water-oxidizing complex (WOC) and the reduction of plastoquinone at the Q_B binding site, producing molecular oxygen, protons and electrons that sustain photosynthetic electron flux [3–5]. It remains the only known enzyme capable of oxidizing water, a process initiated by photon absorption at chlorophyll pigments and stabilized through long-lived charge separation by the reaction center, P680 [2]. Integration of PSII photochemistry with the WOC enables the system to generate both redox products, molecular oxygen and reduced plastoquinol (PQH_2), and a proton gradient across the thylakoid membrane that contributes to the proton motive force that powers ATP synthesis [3, 5]. The catalytic core of the WOC consists of an inorganic Mn_4CaO_5 cluster coordinated within the protein matrix, which accumulates oxidizing equivalents across four charge separation events to drive four-electron oxidation of two water molecules [4]. The operation of PSII is indispensable not only for energy conversion in phototrophs but also for sustaining Earth's oxygenic atmosphere. Despite phototroph phylogenetic and ecological diversity, the inorganic core and protein architecture of PSII remain conserved.

Structurally, PSII is a large membrane protein complex with a highly conserved core of subunits surrounded by less essential and sometimes transient proteins (**Figure 1**). The dimeric holoenzyme contains over 30 protein subunits together with a network of pigments, lipids, and inorganic cofactors. The reaction center is formed by the D1 (PsbA) and D2 (PsbD) proteins, which coordinate the primary electron-transfer cofactors, including the special pair of chlorophylls (P680), pheophytins, plastoquinones at the Q_A and Q_B sites, and the nonheme iron (NHI) [6]. The core antenna protein subunits CP43 and CP47 (PsbC and PsbB) coordinate over 30 chlorophyll molecules and carotenoids that funnel excitation energy into the reaction center [7]. Several low-molecular-mass subunits and extrinsic lumenal proteins further stabilize the complex and regulate water oxidation [8]. Among these, PsbO is conserved as the manganese stabilizing protein, whereas other extrinsic components

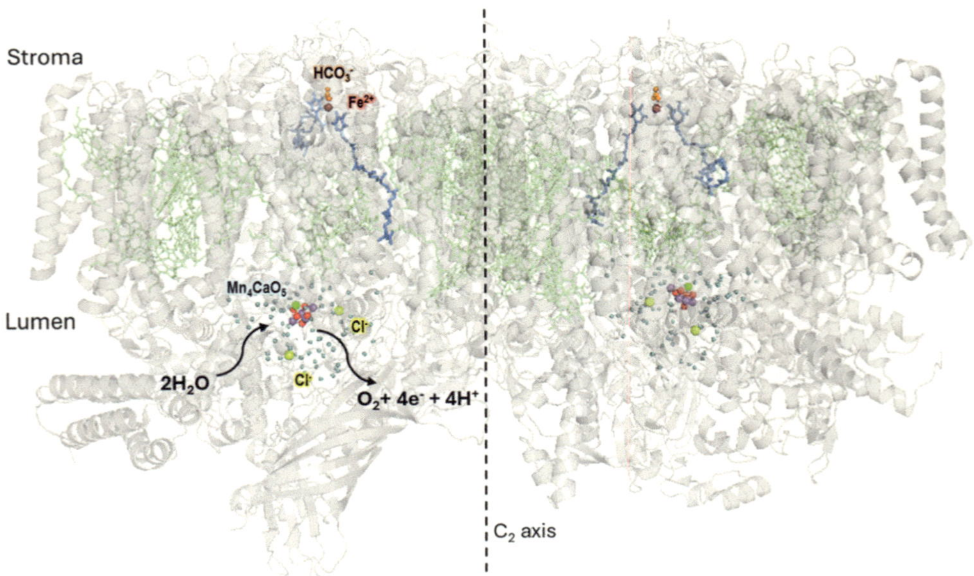

Figure 1.
PSII structure highlighting known positions of relevant cofactors, including the WOC with CA (green), MN1-4 (purple), and bridging oxo ligands (red); two donor-side chloride ions (yellow), the NHI (ruby), and its bicarbonate (orange). Chlorophylls are shown in light green, plastoquinone acceptors in dark blue, and water molecules in cyan. Structure from PDB 78no.

vary by lineage [9]. High-resolution structures of cyanobacterial PSII from *Thermosynechococcus elongatus* and *Synechocystis sp.* PCC 6803 have revealed the spatial organization of protein subunits, pigments, and the Mn_4CaO_5 cluster [10–12]. Structures resolved to near-atomic detail by Xray crystallography and cryo-electron microscopy provide the foundation for understanding how PSII integrates light harvesting, electron transfer, proton translocation, and catalysis, illuminating how protein architecture supports the chemistry of water oxidation [13–15].

While the protein subunits of PSII provide the structural framework, its catalytic activity depends on a precisely tuned network of inorganic cofactors (**Figure 1**). On the donor side, the Mn_4CaO_5 WOC catalyzes the four-electron oxidation of water to O_2 through the Kok S-state cycle, advancing through five metastable oxidation states (S-states, $S_0 \rightarrow S_1 \rightarrow S_2 \rightarrow S_3 \rightarrow [S_4]$) that sequentially accumulate oxidizing equivalents on the redox-active manganese until O–O bond formation and oxygen release occur [5, 16]. Calcium (Ca^{2+}) plays a critical role in stabilizing the cluster geometry and facilitating substrate water binding, while a hydrogen-bonded network of waters and residues coordinates proton removal [17]. Chloride (Cl^-) ions further contribute by maintaining the integrity of this network and enabling efficient proton egress [18]. On the acceptor side, bicarbonate (HCO_3^-) bound near the NHI modulates quinone redox potential, stabilizes forward electron transfer, and minimizes charge recombination pathways [19–21]. Collectively, these cofactors enable PSII to accomplish efficient water oxidation, maintain redox poise, and minimize photodamage.

1.1 Significance

Despite decades of study, the precise mechanistic contributions of these cofactors remain under active investigation. Herein we examine the roles of five inorganic cofactors in PSII: manganese, calcium, chloride, bicarbonate and the NHI (**Table 1**). We emphasize their operational roles in catalysis, electron and proton transfer, and structural stabilization rather than in photoassembly, which has been covered in detail in previous reviews [17, 22, 23]. As presently stands, major questions remain regarding the functions of all of these inorganic cofactors. Therefore, their roles must be elucidated to provide a strong foundation for future research efforts to improve the efficiency of PSII in natural systems, a critical objective for improving crop and biofuel productivity, or develop efficient biomimetic water-oxidizing systems,

Cofactor	Thylakoid position	Role
Manganese	Lumen (WOC)	Redox cycling, water oxidation
Calcium	Lumen (WOC)	Tuning of manganese operation
Chloride	Lumen (near WOC)	Proton removal from the WOC
Bicarbonate	Both sides (NHI)	Mediates protonation, electron-transfer Q_A–Q_B, and WOC photoassembly
Iron	Stroma (Q_A–Q_B)	Mediates electron-transfer Q_A–Q_B

Table 1.
List of inorganic cofactors of PSII, their locations, and roles.

a longstanding goal in renewable energy technology. In the following sections, we discuss each cofactor individually, emphasizing structure of the ligand environment, mechanistic role in PSII, and remaining questions on operation. By integrating structural and mechanistic perspectives, this chapter aims to provide a clear framework for understanding what is known and what is not presently known about how these cofactors regulate water oxidation and maintain PSII function.

2. Manganese

2.1 Structural location

The WOC of PSII comprises the Mn_4CaO_5 cluster, an inorganic cofactor that drives light-driven water oxidation (**Figure 2**). Embedded in a well-defined pocket on the lumenal side of the complex, the cluster adopts a cubane-like geometry wherein three manganese ions (Mn1–Mn3) and a single calcium ion form a heterocubane core, while a fourth "dangler" manganese (Mn4) is linked to the others through μ-oxo bridges [10]. Four terminal waters (W1–W4) coordinate to Mn4 and Ca^{2+} and are essential for O_2 evolution. The cluster is ligated primarily by D1 subunit residues including D1-D170, D1-E189, D1-E333, D1-D342, D1-A344, D1-H337, and D1-H332, along with CP43-E341 (**Figure 2A**). Several of these carboxylates are bridging ligands to adjacent Mn and Ca^{2+}, while D1-H332 and D1-E189 ligate Mn1 in a monodentate manner [24]. Notably, D1-D170 provides a high-affinity site for the first Mn^{2+} and oxidation during the photoassembly of the WOC [25]. These μ-oxo bridges, terminal ligands, and carboxylate coordination create a stable yet redox-flexible framework that stabilizes high-valent manganese while allowing storage of multiple oxidizing equivalents.

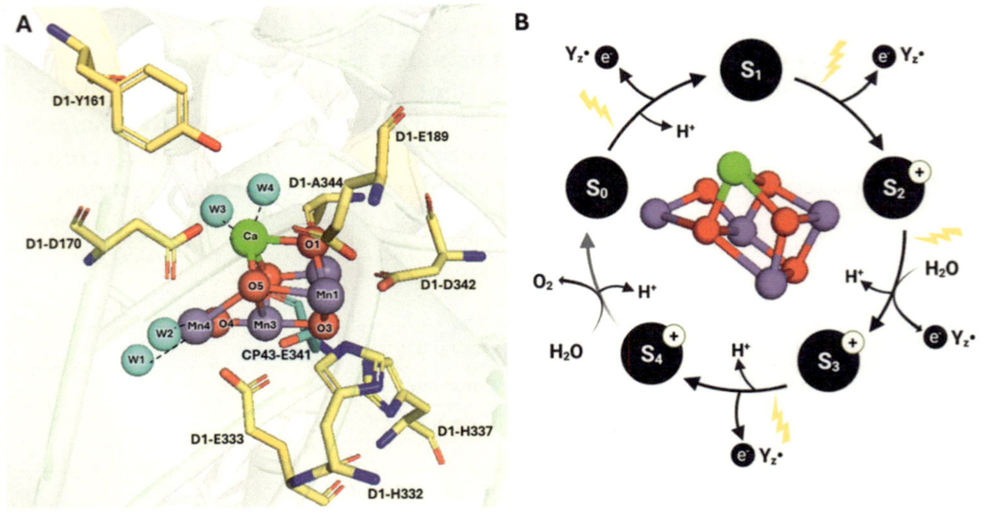

Figure 2.
(A) The WOC of PSII, showing the Mn_4CaO_5 cluster with conventional numerical assignments of atoms, the four conserved water ligands to Mn4 and Ca, and the ligands of the first coordination sphere, as well as tyrosine-z for directional reference. (B) The WOC cycle of S-state transitions showing oxidation, deprotonation, and water uptake events.

High-resolution crystallography revealed the geometry of the cluster as a Mn_3CaO_4 pseudocubane linked to a terminal manganese ion [2, 10]. These breakthrough structures established the foundation for current models but have well-known limitations: The manganese ions are prone to reduction under synchrotron Xrays, the exact positions of μ-oxo and water ligands remain difficult to resolve because oxygen scatters weakly relative to the surrounding metals, and the protonation states of the oxygens within the WOC remain unknown, being unobservable with current approaches [15, 26, 27]. Complementary spectroscopic methods have provided key constraints: Extended Xray absorption fine structure demonstrated metal–metal and metal–ligand distances, while electron paramagnetic resonance (EPR) and ENDOR spectroscopy assigned manganese spin states across the Kok cycle [28, 29]. Computational studies using DFT and QM/MM integrate experimental results to generate several models that do not completely align [30, 31].

2.2 Redox cycling and catalytic function

The functional role of Mn in PSII is defined by its ability to undergo sequential one-electron redox transitions. Within the Mn_4CaO_5 cluster, the Mn ions cycle primarily between the +3 and +4 oxidation states, allowing stepwise accumulation of four oxidizing equivalents required for water splitting [28]. Each charge separation event generates the strong oxidant $P680^+$, which is quickly reduced by the redox-active tyrosine TyrZ (D1-Y161) [32]. Upon donating an electron, TyrZ forms a neutral radical (TyrZ·) which transfers the oxidizing equivalent to the manganese cluster through the hydrogen-bonded D1-H190 network [16]. This rapid sequence ensures the oxidizing power generated by light absorption is stabilized and stored at the catalytic site. Each manganese oxidation step is coupled to proton release from μ-oxo or water ligands, preventing excessive charge accumulation and maintaining the energetic balance [33].

2.3 Manganese in the S-state transitions

The four manganese ions in the Mn_4CaO_5 cluster sequentially accumulate the oxidizing equivalents required for water splitting. Progression through S-state transitions drives stepwise electron removal, proton release, and water substrate activation, ultimately enabling O=O bond formation [34, 35]. Although it is known that the S-states advance through successive light driven charge separation events, the exact electronic configuration of the metal ions at each step remains under investigation. The high covalency and charge delocalization of the cluster complicates oxidation state assignments, and while spectroscopic, crystallographic, and computational studies have provided insights, they also highlight areas where interpretations remain unresolved [13, 28]. Early spectroscopic work demonstrated that each S-state transition is linked to electron removal and proton release, but subsequent studies indicated that proton release is not uniform across the cycle, producing a characteristic 1:0:1:2 pattern of proton stoichiometry (starting from the $S_0 \rightarrow S_1$ transition) [36, 37].

The $S_0 \rightarrow S_1$ transition involves the oxidation of the most reduced state of the cluster to the dark stable S_1 state. It is associated with the loss of one electron and one proton, preparing the cofactor for further charge accumulation steps in the cycle. Spectroscopic evidence including EPR and ENDOR studies suggests that the S_1 state is diamagnetic and represents the resting state under physiological conditions [13, 32]. While the exact oxidation configuration of the manganese ions in S_0 and S_1 remains debated across different experimental and computational models [28], there is a broad agreement that this transition establishes the baseline state from which the water oxidation cycle proceeds.

The $S_1 \rightarrow S_2$ transition is initiated by oxidation of the water oxidizing complex via $Y_Z\cdot$, resulting in the removal of an additional electron from the cluster without an associated proton release [34, 38]. This is distinct within the Kok cycle because it is the only step when net charge accumulates on the cofactor. EPR spectroscopy has characterized the S_2 state, revealing two interconvertible spin isomers: a multiline signal at $g \approx 2$ and a broad feature at $g \approx 4.1$ [39, 40]. These signals are commonly associated with "open" and "closed" cubane conformations of the Mn_4CaO_5 cluster [41]. Xray free electron laser (XFEL) studies have provided structural evidence consistent with both configurations, leaving unresolved which conformation predominates under physiological conditions [42, 43]. The existence of multiple electronic and structural isomers emphasizes the flexibility of the S_2 state, which is proposed to enable water insertion and proton release during the successive transition [44].

The transition from $S_2 \rightarrow S_3$ is widely recognized as a key step where the water oxidizing complex undergoes oxidation and structural rearrangement. In addition to electron transfer, this transition is associated with proton release, reflecting changes in the hydrogen-bond network surrounding the Mn_4CaO_5 cluster [34]. Structural studies using femtosecond Xray crystallography have reported an additional oxygen ligand (O6), in the S_3 state [42, 43, 45, 46]. Its positioning near O5 led to proposals that O6 could function as a substrate oxygen. Early structural models suggested a short O5–O6 distance, posited as O–O bond formation [45]. However, high-resolution studies have reviewed this distance, indicating that O–O formation has not occurred at this stage [43, 47]. Various models describe O6 as a ligand bound to manganese, as a water molecule from the Ca^{2+} site, or as a preloaded substrate for the next cycle [4, 48]. While the presence and function of O6 is uncertain, the $S_2 \rightarrow S_3$ transition is recognized as a step for structural and electronic reorganization that prepares the cluster for oxygen evolution.

The final $S_3 \rightarrow [S_4] \rightarrow S_0$ transition is when O=O bond formation and the release of two protons and O_2 occur, with a reset of the Mn_4CaO_5 cluster. The S_4 state itself is a highly transient intermediate, detected only indirectly through time-resolved spectroscopy and kinetic studies, which has made its electronic characterization difficult [35, 43]. Time-resolved Xray spectroscopy and fluorescence studies have suggested that after Y_z oxidation, an intermediate state persists for hundreds of microseconds before O_2 release, a period associated with a proton release [38, 49]. The exact electronic configuration of this intermediate remains debated: Some mechanistic models propose a Mn-oxyl species that undergoes oxo–oxyl coupling, while other groups propose a formation of a high-valent Mn-oxo intermediate that undergoes nucleophilic attack by bound hydroxide or water [50]. In both cases, structural studies suggest O5 as one substrate oxygen, while the second may correspond to a newly inserted ligand such as O6 or a calcium-bound water seen in XFEL structures [14, 42, 43]. While the exact pathway remains unknown, it is agreed that that proton releases occur both before and after O–O bond formation and that cluster reorganization following O_2 release restores the S_0 state.

2.4 Coupling of proton and electron transfer

Manganese oxidation within the Mn_4CaO_5 cluster modulates the proton affinity of μ-oxo bridges and terminal waters, shifting the local pKa and coordinating proton release coupled to electron removal. This coupling gives rise to the characteristic 1:0:1:2 proton release pattern observed in the Kok cycle, preventing excessive charge buildup and maintaining redox potentials accessible to Y_z·/Y_z [5, 34]. Proton transfer is gated by Y_z oxidation and directed through hydrogen-bond networks such as the Cl^- channel, ensuring each electron transfer to $P680^+$ proceeds in favorable electrochemical conditions and maintains synchronicity of charge accumulation and substrate activation [28, 36]. Studies from site-directed mutagenesis, including D1–D61 and CP43-R357 variants, revealed that disruption of these networks leads to slowed S-state transitions and altered proton release, connecting residues to catalytic gating [51, 52]. Metal substitution studies further emphasized the functional distinctiveness of manganese: Substitution with cobalt disturbs charge accumulation and delays the Kok cycle before O_2 evolution, underlining the requirement for manganese's redox flexibility to support multielectron chemistry [53]. PSII has evolved a finely tuned network in which manganese oxidation, proton transfer, and Y_z-mediated electron flow are coordinated to accomplish efficient water splitting.

3. Calcium

3.1 Structural description

Within the Mn_4CaO_5 cluster of PSII, calcium occupies the corner of the WOC distorted cubane opposite Mn3 [10, 26]. The pseudocubane is not static; the cluster can exist in either an "open" or "closed" configuration in at least some S-states [54]. In the open cubane, Mn4 is connected to the core by O4 and O5, while in the closed cubane, Mn4 solely ligates O4. The open form is believed to facilitate substrate water access and O–O bond formation by maintaining Mn4 in a reactive position [28]. Calcium stabilizes the open cubane geometry through its specific coordination pattern, enforcing the arrangement for efficient catalysis. Calcium is oxo-bridged to all 4 Mn ions via di-μ-oxo bridges with Mn1 via O1 and O5, Mn2 via O1 and O2, and Mn3 via O2 and O5; and a mono-μ-oxo bridge with Mn4 via O5. Calcium also coordinates waters 3 and 4.

3.2 Function of calcium

Calcium is essential for O–O bond formation because it stabilizes high-valent Mn intermediates and provides a ligated water molecule (either W3 or W4) to potentially couple with a Mn-oxo species [55]. Calcium maintains Mn–Mn distances within a catalytically suitable range and tunes the electronic environment to facilitate proton-coupled electron transfer [56]. By sharing μ-oxo ligands with Mn ions, calcium indirectly tunes their redox potentials by lowering the energetic cost of reaching high-valent states necessary for O–O bond formation and facilitating the $S_2 \rightarrow S_3$ transition [57], as observed via FTIR studies [58]. Calcium's electrostatic field is further proposed to stabilize reactive Mn-oxo intermediates by polarizing μ-oxo ligands and enabling delocalization of charge, which stabilizes Mn(IV) intermediates in $S_2 \rightarrow S_3 \rightarrow S_4$ transition states and prevents destructive radical accumulation [32].

3.3 Atomic substitution

Substitution studies demonstrate the unique role of calcium in the structure and reactivity of the WOC. Sr^{2+} is the only ion able to replace calcium while supporting oxygen evolution [59]. Crystallographic studies demonstrate that Sr^{2+} occupies the calcium WOC site, preserving the geometry of the cluster despite its larger ionic radius [60]. Sr^{2+} substitution results in slower O–O bond formation, reduced quantum efficiency, and altered S-state lifetimes [61]. Sr^{2+} influences the WOC toward a closed cubane geometry [62]. All other divalent cations cannot replicate these functions. Cd^{2+} can be incorporated but blocks advancement through the S-states beyond the S_2-state and distorts Mn–Mn distances [63], while no other substituent allows WOC turnover or S-state advance. Disruption of the calcium site interferes with the linkage of substrate delivery at W3/W4 to the timing of proton and electron-transfer events [58].

3.4 Remaining questions

Several aspects of calcium's function in the water oxidizing complex remain unknown despite decades of study. One outstanding question is whether calcium, as a Lewis acid, can redistribute electron density within the cluster. Calcium is proposed to displace electron density across the Mn cluster, modulating reactivity [64]. Biochemical studies indicate that incorporation of calcium guides manganese binding and oxidation in photoassembly. Omission or substitution of calcium stalls this process at intermediate states [65]. During PSII repair, partial turnover of calcium has been observed, consistent with its high stoichiometric requirement in photoassembly [66]. In addition to the catalytic site, a secondary calcium-binding site has been identified in the extrinsic PsbO subunit, which does not participate in O–O bond formation [67]. Removal of calcium from this peripheral calcium-binding pocket weakens PsbO binding, which indirectly destabilizes the WOC and reduces oxygen evolution efficiency without affecting the primary catalytic mechanisms. Mechanistically, this site may help maintain the structural integrity of loop regions that shield the WOC and regulate substrate water accessibility. It may also act as a local reservoir for calcium during photoassembly and repair cycles.

4. Chloride

Beyond metal cofactors, small anions play critical roles in PSII function. Among these, chloride has received attention due to its essential role in supporting water oxidation and stabilizing the water oxidizing complex during both catalysis and assembly. The following section summarizes the current understanding of chloride's binding sites, mechanistic roles, and impact on PSII efficiency.

4.1 Location and binding sites

Structural, spectroscopic, and computational approaches have established that PSII contains two conserved chloride-binding sites per monomer, located on the lumenal side near the Mn_4CaO_5 cluster. Early studies showed that Cl^- depletion inhibits S-state transitions and revealed a strict Cl^- requirement for O_2 evolution. The precise locations of these ions near the WOC were determined by Xray crystallography and reinforced by cryo-EM [10, 11]. The two sites, known as Cl1 and Cl2, are located on opposite sides of the Mn_4CaO_5 cluster, approximately at >7 Å from the nearest manganese ion and ~11 Å from the calcium ion, within hydrogen-bond networks of key residues but outside the first coordination sphere (**Figure 3**) [68, 69]. Cl1 is located at the entrance of the broad (Cl1) water channel leading to the lumen and is surrounded by the side chains of D1-N181, D2-K317, and nearby D1–D61 [70]. Cl2 lies in a small pocket adjacent to the WOC near the narrow O4 channel, coordinated by the backbone of D1-N338 near the hydrogen-bond network [10]. Both sites are adjacent to ordered water networks that connect the catalytic cluster to the lumen [71]. While occupancy can vary depending on sample preparation, the binding sites are consistently observed across cyanobacterial and higher plant PSII structures [10, 11].

4.2 Functional roles

Cl^- is essential for sustaining efficient PSII turnover, particularly by facilitating proton removal from the Mn_4CaO_5 cluster to the lumen during the $S_2 \rightarrow S_3$ and $S_3 \rightarrow S_0$ transitions [72–74]. Cl^- depletion slows these transitions, leading to impaired O_2 evolution and accumulation of oxidized intermediates. Cl^- also helps to maintain the

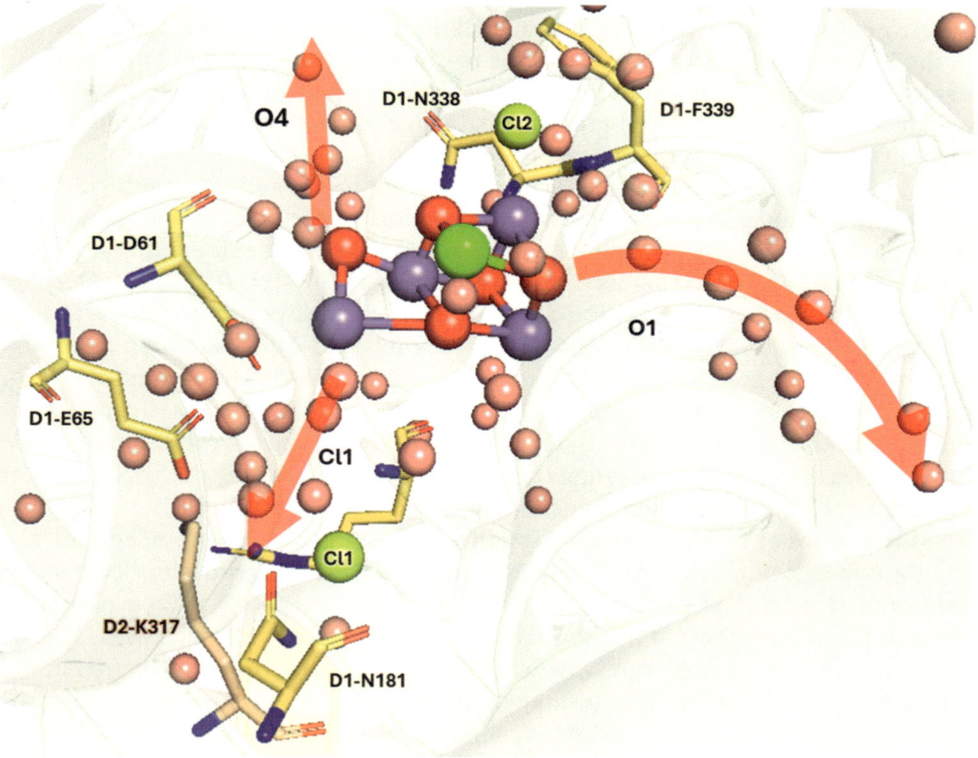

Figure 3.
Close-up view of the canonical chloride-binding sites in PSII, highlighting both Cl1 and Cl2 positions (yellow). Surrounding channels (red) and residues from D1 (yellow) and D2 (beige) that interact with each site are shown, with hydrogen-bonding waters (salmon) displayed as spheres.

proper electrostatic environment near the catalytic site. The Cl1 site modulates the protonation states of nearby residues such as D1–D61 and stabilizes nearby hydrogen-bond networks [18]. Occupancy of this site prevents the formation of an inhibitory salt bridge between D1–D61 and D2-K317, which would block the proton egress pathway channel and halt catalysis [69, 75, 76]. Some studies further suggest that Cl^- contributes to stabilizing the Mn_4CaO_5 cluster and its surrounding environment, though the exact mechanism remains debated [69, 77, 78]. Evidence from halide substitution studies suggests that chloride's role is not purely structural but chemically specific and optimized for PSII, as anion substitution leads to impaired proton removal and decreases photosynthetic efficiency [73, 79]. Through these electrostatic and structural contribution effects, Cl^- acts as a gatekeeper of proton transfer.

4.3 Mechanistic evidence

Several studies have established a relationship between Cl^- binding, generally at the Cl1 site, and efficient PSII catalysis. Early chloride depletion and reconstitution experiments showed that removing Cl^- from isolated PSII membranes significantly reduces O_2 evolving activity and delays the WOC cycle at the $S_2 \rightarrow S_3$ and $S_3 \rightarrow S_0$ transitions where proton release occurs [80]. Readdition of chloride restores both S-state advancement and O_2 evolution, indicating that the effect is reversible and reflects a specific requirement for chloride rather than irreversible damage. Spectroscopic studies, including EPR and FTIR, reveal characteristic modifications in the S_2 multiline signal and shifts in the hydrogen-bonding network under Cl^- depletion [78, 81]. In addition, time-resolved fluorescence and flash-induced O_2 evolution have shown delayed proton release kinetics and increased charge recombination in the absence of Cl^- [68, 80].

Additional mechanistic insights have come from halide and anion substitution studies, which investigated whether Cl^- is chemically specific or simply structural. Bromide is the most studied replacement because of its similar size and charge, with crystal structures confirming its occupancy at the Cl1 site [69, 70, 73, 82]. Despite this, bromide slows the $S_2 \rightarrow S_3$ and $S_3 \rightarrow S_0$ transitions, delays proton release to the lumen, reduces and slows the formation of the thylakoid ΔpH, and increases charge recombination at moderate-to-high light intensities [82–84]. These effects demonstrate that Cl^- optimizes proton-coupled electron-transfer kinetics and minimizes recombination [84, 85]. Iodide can partially rescue O_2 evolution but slows proton release and promotes competing side reactions, including sacrificial electron donation and Y_Z modification, which disrupts electron flow [68, 83].

Mutagenesis of residues surrounding the Cl1 binding site has provided critical insight into chloride's functional role. Mutants of D2-K317 disrupt the electrostatic environment near the Cl1 channel and reduce chloride affinity, yet O_2 evolution remains partially supported [76, 86]. This lysine can form a salt bridge with D1–D61, which would block proton egress, when chloride is absent [18, 76]. In contrast, mutations at D1-N181, another residue near Cl1, do not eliminate the binding of Cl^- but lead to significantly reduced steady-state O_2 evolution and slower O–O bond formation despite normal S-state cycling [87]. These results imply that D1-N181 tunes the geometry of water ligands and the hydrogen-bond network rather than acting as a gate in proton transfer. Altogether, mutagenesis studies suggest that Cl1 acts as an electrostatic cofactor that stabilizes the hydrogen-bond network around the Mn_4CaO_5 complex and helps prevent inhibitory ion-pair formation, ensuring efficient turnover under physiological conditions [18].

Beyond steady-state catalysis, studies suggest that chloride plays an important role during photoassembly of the Mn_4CaO_5 cluster, particularly in the early light-driven steps when Mn^{2+} is oxidized to Mn^{3+} and proton removal is required. Cl^- addition to manganese-depleted PSII membranes accelerates photoactivation, shortens the lag phase, and more effectively restores O_2 evolution [88, 89]. Under low chloride conditions, an inverse H/D solvent isotope effect has been observed, indicating that proton removal at the high-affinity manganese-binding site becomes rate-limiting and that chloride reduces this barrier [66]. Cl^- is also suggested to promote Mn^{3+} formation during the first oxidation step, likely by stabilizing the hydrogen-bond network and enabling deprotonation events required to form μ-oxo/μ-hydroxo bridges [88]. Models also suggest that chloride's presence modulates the local proton environment to facilitate these early deprotonation events, though the molecular mechanism remains under investigation [90].

4.4 Water channel function

Within PSII, several water-filled channels connect the Mn_4CaO_5 cluster to the lumen, including the Cl1 (broad) and O4 (narrow) channels, which have been proposed as major proton egress routes during S-state cycling based on crystallography and computational studies [91, 92]. These channels are partially hydrophobic pathways lined with protonatable residues and ordered waters, forming a highly organized hydrogen-bond network that appears to enable proton transfer through Grotthuss mechanisms while limiting bulk solvent intrusion [91–93]. The Cl1 channel leads past the mentioned Cl1 site between D1–D61 and D2-K317 toward the lumen, positioning chloride as a key regulator of channel integrity and modulating the protonation states of nearby residues to lower the energetic barrier for proton translocation [10, 11, 18, 69, 70, 85]. In this role, Cl^- synchronizes proton egress with the $S_2 \rightarrow S_3$ and $S_3 \rightarrow S_0$ transitions, sustaining efficient O_2 evolution [18]. Spectroscopic, computational, and mutagenesis studies determined Cl1 to be the primary proton egress pathway during these transitions, while the O4 channel is sometimes associated with proton release during the $S_0 \rightarrow S_1$ transition, varying by organism and PSII preparation [85, 91, 94]. Cl^--dependent proton transfer was proposed to work alongside transient bicarbonate buffering at arginines near the Cl1 channel to neutralize excess charge, maintain electrostatic balance, and limit recombination under high-flux conditions [84, 95]. Although the precise role of Cl2 remains less defined, it is consistently observed in a pocket adjacent to the O4 channel [10, 70]. It has been proposed to help stabilize the hydrogen-bond network in this region, potentially influencing proton release in the early S-state transitions, but direct mechanistic evidence is still limited and its role remains an open question [18, 69].

4.5 Outstanding questions

Despite advances in understanding chloride's function in PSII, several key questions remain. Although Cl1 is widely accepted as a key proton egress site, the timing of proton release and its relationship to alternate routes such as the O4 channel are still debated. The functional significance of Cl2 is still unclear, with only limited evidence for its involvement in proton transfer or water delivery. Resolving these roles will be essential for establishing a mechanism for proton management on the donor side of PSII.

5. Bicarbonate

5.1 Historical context

Among the inorganic cofactors of PSII, bicarbonate (HCO_3^-) has drawn attention because of its conserved binding site at the acceptor side and its still-debated mechanistic roles in facilitating electron transfer and proton management. Bicarbonate exists in equilibrium with dissolved CO_2 and plays both structural and regulatory roles [20, 96, 97]. It was first proposed as a direct participant on the donor side of PSII [98, 99], but while depletion experiments revealed photoinhibition and impaired electron transport [100, 101], the precise site and mechanism of action remained unclear. Govindjee and collaborators established the "bicarbonate effect" within PSII research [102]. Today, bicarbonate is firmly associated with the NHI (**Figure 4**) on the acceptor side of PSII [103–105], although donor side contributions have also been proposed [106], and several groups continue to actively research its role.

5.2 Acceptor side

5.2.1 Structural location

Bicarbonate is localized to the NHI on the acceptor side of PSII, positioned between the primary quinone electron acceptor Q_A and the secondary, exchangeable

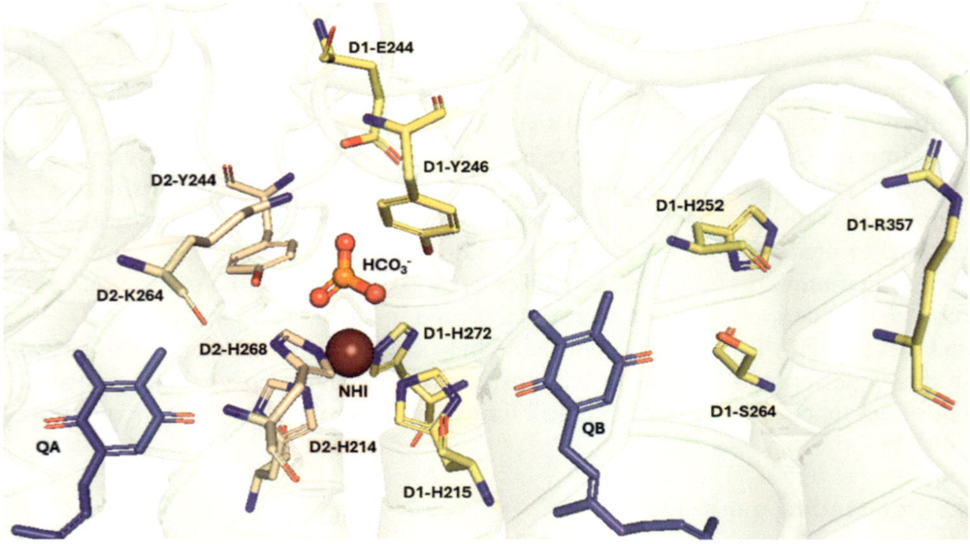

Figure 4.
The acceptor side of PSII, showing the quinones Q_A and Q_B (blue), the NHI (ruby) situated between them, and its coordinated bicarbonate ion (orange). Key residues from the D1 (light yellow) and D2 (beige) subunits which directly coordinate or interact with the NHI and the canonical bicarbonate site are shown. Additional residues relevant to protonation of the distal carbonyl of Q_B are shown at right.

quinone Q_B. Early spectroscopic studies provided indirect evidence for this placement. EPR signals from Q_A^- near the nonheme Fe^{2+} are perturbed by formate addition, suggesting that bicarbonate shares the same coordination sphere at the NHI [107]. FTIR spectroscopy revealed signatures of bicarbonate ligation, suggesting monodentate and bidentate coordination modes that depend on the oxidation state of the NHI [108]. Structural confirmation came from high-resolution crystallography of cyanobacterial PSII at 1.9 Å, which was followed by observations across cyanobacteria, algae, and higher plants, which consistently revealed bicarbonate coordinating to the NHI as a bidentate ligand [10, 11, 49, 75, 109, 110]. Nearby residues such as D1-R257 and D1-H252 may further anchor the bicarbonate through hydrogen bonding [111]. Further stabilization is provided by D2-Y244 and D1-Y246, which extend a hydrogen-bonding network and maintain the geometry of the bicarbonate ligand [10, 112]. Structural analysis of PSII assembly intermediates suggests that bicarbonate binding is dynamic during biogenesis. In partially assembled complexes, the NHI can instead be ligated by a glutamate from the D2 subunit [113, 114], but in the fully assembled PSII complex, this position is occupied by bicarbonate.

5.2.2 Functional roles

Beyond its structural placement at the NHI between Q_A and Q_B, bicarbonate fulfills several roles essential for PSII function. By coordinating with the NHI, bicarbonate stabilizes electron transfer between the primary and secondary quinones and fine-tunes redox potential at this site [19, 108, 115]. This coordination enhances efficiency of forward electron flow and minimizes charge recombination between $P680^+$ and Q_A^- [116]. In the absence of bicarbonate, O_2 can occupy this site and accept electrons, leading to superoxide formation. Although this provides a photoprotective outlet, it is less efficient and increases the risk of reactive oxygen species (ROS) damage [106, 117]. In addition, bicarbonate has been proposed to facilitate the protonation of Q_B to PQH_2, although the extent of this contribution remains under investigation [118].

5.2.3 Mechanistic evidence

Early functional experiments provided the first mechanistic clues about bicarbonate's role on the acceptor side of PSII. Loss of HCO_3^- by CO_2 removal or displacement with formate consistently slowed Q_A^- reoxidation and impaired plastoquinone reduction, establishing a direct link between bicarbonate availability and electron-transfer efficiency [21, 115, 119]. Fluorescence decay patterns under HCO_3^- depletion resemble DCMU inhibition, indicating that the block was after Q_A but prior to the entry into the PQ pool [119, 120]. Herbicide-binding studies further revealed that atrazine affinity decreased under bicarbonate depletion, implicating its role at or near the Q_B site [120, 121]. Restoration of activity upon reintroduction of bicarbonate and partial rescue with small anions such as formate or acetate reinforced the view that the bicarbonate ligand at the NHI is directly coupled to the kinetics of Q_A to Q_B electron transfer [107]. Bicarbonate binding lowers the midpoint potential of Q_A/Q_A^-, favoring downstream electron transfer to Q_B and suppressing recombination with $P680^+$. Loss of bicarbonate shifts this potential to more positive values, slows Q_A^- to Q_B transfer, and increases the probability of charge recombination, explaining elevated photodamage under bicarbonate depletion [116, 122].

Formate replacement introduces a pH-dependent conformational change that reduces protonation efficiency at Q_B, suggesting bicarbonate's dual role in electron transfer and proton delivery [123]. Anion substitution experiments further demonstrated that while ligands such as formate can coordinate the NHI, they fail to reproduce native electron-transfer kinetics, indicating that bicarbonate uniquely optimizes the energetics for plastoquinone reduction [20, 100, 105, 106]. Structural modeling and mutagenesis studies have suggested that residues near D1-H215, D1-H252, and D1-S264 form critical nodes of this pathway, with D1-Y246 providing an H-bond bridge between the HCO_3^- ligand and the Q_B pocket [124, 125]. In *Limnospira maxima*, depletion allows the first electron transfer to Q_B but blocks PQH_2 formation, associating either the NHI or the stromal surface near D1-H252 and D1-S264 as critical proton delivery sites [126]. This suggests that bicarbonate's role is not only redox tuning but also direct coordination of proton-coupled electron transfer for efficient PQH_2 turnover and photoprotection [127].

5.3 Donor-side effects

5.3.1 Structural role and location

While bicarbonate is well established as a ligand to the NHI on the acceptor side of PSII, its structural role on the donor side remains unresolved. In contrast to chloride, which occupies defined proton egress channels, high-resolution structural studies showed no evidence of a fixed bicarbonate binding site on the donor side [11, 75, 109, 110]. Instead, biochemical and spectroscopic studies suggest more transient or dynamic associations. Several studies support the idea that bicarbonate can associate dynamically with positively charged residues, particularly arginine and lysine, throughout PSII. Notably, CP43-R357, located in the second coordination shell of the WOC, has been highlighted as a potential interaction site. Its substitution leads to a decrease in O_2 evolution dependent on bicarbonate concentration [52, 128].

5.3.2 Functional and mechanistic roles during catalysis

Although no stable donor-side binding site for bicarbonate has been resolved structurally, functional studies suggest a mechanistic contribution to PSII catalysis [96, 129]. Bicarbonate may serve as a mobile proton carrier and buffer, diffusing through hydrogen-bond networks to neutralize protons released during water oxidation [100, 130]. This role complements that of Cl^-, which helps define static proton egress pathways, whereas bicarbonate extends buffering capacity under conditions of high proton release rates or changed hydrogen bonding [66, 89]. HCO_3^- restoration of normal S-state cycling in the D2-K317A mutant but not in D1-D61A or CP43-R357K supports the idea that proton removal is mediated by a diffusible base such as bicarbonate rather than a fixed residue [128]. Bicarbonate improves O_2 evolution under high light and rescues activity in hydrogen-bond network mutants [100, 130]. Beyond its catalytic role, HCO_3^- has been linked to photoassembly of the Mn_4CaO_5 cluster. During photoactivation, stepwise addition of Mn^{2+} ions and oxidation are coupled to proton release, requiring local buffering to stabilize the high energy intermediate states [131, 132]. Bicarbonate is proposed to associate with positively charged residues near the donor side, electrostatically steering Mn^{2+} toward the apo-WOC site and thus enhancing yield and rate of assembly [133]. Isotopic tracing using ^{13}C-labeled bicarbonate provided direct evidence of transient coordination to Mn^{2+} during the initial photoassembly steps, stabilizing Mn^{3+} and facilitating succeeding metal ion incorporation [134]. Depletion of bicarbonate inhibits O_2 evolution recovery after photodamage, while reinserting bicarbonate reestablishes efficient photoactivation [106]. Together, these findings indicate that bicarbonate serves dual roles on the donor side: (1) buffering of protons during water oxidation to prevent charge accumulation; and (2) enabling photoassembly and repair of the WOC by stabilizing Mn intermediates and promoting their correct positioning.

5.4 Outstanding questions

Although bicarbonate's role at the NHI is well known, uncertainties remain regarding whether it behaves exclusively as a bound cofactor or also as a mobile, diffusible cofactor. Functional studies suggest that bicarbonate can buffer WOC proton release through hydrogen-bond networks [100, 130, 131], but structural evidence has localized it to the acceptor site [10, 11]. Species variation adds complexity; for instance, *Limnospira maxima* lives in extremely high concentrations of bicarbonate and exhibits unusual PSII behavior, raising the possibility that bicarbonate plays a distinct mechanistic role in an adapted organism [106, 128]. Structurally, donor-side effects remain unresolved: No structure has resolved bicarbonate near the WOC [10, 11, 110], and standard purification protocols may unintentionally deplete weakly bound bicarbonate [99]. While studies have associated bicarbonate in photoassembly and repair of the Mn_4CaO_5 cluster [23, 132], current evidence is restricted by assays and the inability to determine short-lived intermediates. Addressing these limitations requires integrative approaches combining structures, spectroscopy, and organism-specific studies to elucidate how bicarbonate supports catalysis and biogenesis.

6. Nonheme iron

6.1 Structure and function overview

A mononuclear NHI is located between the Q_A and Q_B quinones on the acceptor side of PSII. NHI is a hexacoordinate Fe^{2+} that has a distorted octahedral geometry, coordinating the bicarbonate ion discussed in the previous section and four histidines: D1-H215, D1-H272, D2-H214, and D2-H268. D2-H214 and D1-H215 are positioned between the NHI and Q_A and Q_B, respectively, forming hydrogen bonds to their respective quinones [10, 19]. Each of these four histidines interacts via an unprotonated nitrogen in the imidazole ring as a monodentate ligand, while the structural bicarbonate stromal to the NHI is a bidentate ligand via two of its oxygens [19]. NHI mediates electron transfer between Q_A and Q_B, but does not normally play a direct redox role [19]. However, strong oxidants may convert NHI to iron(III) [135], which then accepts electrons from Q_A^- to return to ground state, leading to observations of two-electron removal from the WOC on a single flash [136]. This phenomenon appears to be limited to conditions under which the structural bicarbonate ligand or similar anions are unavailable [137]. NHI's redox midpoint potential is relatively high, ~360 mV at pH 7.8 in the PSII reaction center [138]. Studies showing pH dependence suggest coupling to nearby protonatable residues [139]. NHI's presence is needed to maintain geometry and electrostatic environment between cofactors including pheophytin and Q_A [19]. NHI has been posited to play a role in ROS detoxification when redox-active [140]. Removal of Fe^{2+} from *T. elongatus* PSII does not impede electron transfer from pheophytin to Q_A [141], excluding a role of NHI at this step. However, EPR and ENDOR spectroscopy following Fe^{2+} depletion [142] determined that Q_A^- is not magnetically coupled to Fe^{2+}. Zn^{2+} has been shown to occupy the NHI site and restore native-like structural, if not functional, features, while copper can similarly mimic the structure but is redox-active in the site [143].

6.2 Role in photoprotection

NHI has historically been viewed as a participant in photoprotective mechanisms. When charge recombination occurs between reduced Q_A^- and oxidized $P680^+$, singlet oxygen, a ROS implicated in photodamage, can be formed as a by-product. Redox potential and stability of Q_A^-, which are regulated by NHI and its bicarbonate ligand, influence recombination probability. The redox potential of NHI decreases across pH 6–8, showing sensitivity to protonation in the environment [138]. Ishikita and Knapp proposed that pH dependence is due to coupling to a network of protonatable residues near NHI that could stabilize Q_A^- and reduce the rate of recombination with $P680^+$ [139]. This buffering effect was proposed to delay harmful back reactions and decrease singlet oxygen formation, especially when forward electron transfer to Q_B is impaired. NHI would thus modulate the flow of electrons. Bicarbonate availability and identity of the D1 protein isoform influence the redox potential of NHI [144]. These may be modulated under stress conditions, since maintaining redox control on the acceptor side is necessary for photoprotection.

6.3 Open questions

While the two-electron gate mechanism of electron transfer through Q_B is established, the protonation steps remain a subject of study. Structural data suggest two-proton delivery pathways, one involving D1-H215 and D1-Y246 for the Q_B carbonyl proximal to the NHI, while the other involves D1-S264 and D1-H252 near the distal site [19]. Bicarbonate stabilizes these networks, but it is not clear if it directly delivers protons [124, 126]. Manipulations on the donor side, such as Ca^{2+} depletion or Sr^{2+} substitution, shift Q_A redox properties. This suggests that there is a long range coupling between Mn/Ca proton dynamics and acceptor-side redox tuning [145]. Such interactions suggest that there may be an integrated electrostatic network, yet the exact mechanistic links are unknown.

7. Conclusion

While the inorganic cofactors of PSII have been known for decades, their functions remain incompletely understood and specific significant questions remain for each. Regarding the manganese and calcium that make up the WOC, targets of future research include the specific mechanism of water incorporation, the composition of the S_4 state, and the precise oxidation states of manganese that are needed to form molecular oxygen. Cl1 is comparatively well understood, but the specific mechanism of proton egress it facilitates remains unknown and there is still no clear evidence for the role of highly conserved Cl2. Bicarbonate's role at the NHI is quite clear, but its donor-side site activity is still unknown and its role in protonation on the acceptor side appears to extend beyond the known NHI site. Finally, the extent of redox activity of the NHI *in vivo* remains to be determined. Obtaining answers to these questions will allow for improved efficiency of conversion of light to human-available energy in a range of applications.

Author details

Leslie Castillo, Nidhi Patel, Nicole Seliga, Gustavo Chavez and Colin Gates*
Loyola University Chicago, Chicago, USA

*Address all correspondence to: cgates4@luc.edu

References

[1] Dismukes GC, et al. The origin of atmospheric oxygen on Earth: The innovation of oxygenic photosynthesis. Proceedings of the National Academy of Sciences. 2001;**98**:2170–2175

[2] Shen J-R. The structure of Photosystem II and the mechanism of water oxidation in photosynthesis. Annual Review of Plant Biology. 2015;**66**:23–48

[3] Cox N, et al. Current understanding of the mechanism of water oxidation in Photosystem II and its relation to XFEL data. Annual Review of Biochemistry. 2020;**89**:795–820

[4] Li H, et al. Oxygen-evolving photosystem II structures during S(1)-S (2)-S(3) transitions. Nature 2024;**626**:670–677

[5] Vinyard DJ, et al. Photosystem II: The reaction center of oxygenic photosynthesis. Annual Review of Biochemistry. 2013;**82**:577–606

[6] Barber J, Nield J. Organization of transmembrane helices in photosystem II: Comparison of plants and cyanobacteria. Philosophical Transactions of the Royal Society of London. Series B: Biological Sciences. 2002;**357**(1426): 1329–1335. discussion 1335, 1367

[7] Barber J, et al. Revealing the structure of the photosystem II chlorophyll binding proteins, CP43 and CP47. Biochimica Et Biophysica Acta (BBA) - Bioenergetics 2000;**1459**: 239–247

[8] Heinz S, et al. Analysis of photosystem II biogenesis in cyanobacteria. Biochimica Et Biophysica Acta (BBA) – Bioenergetics. 2016;**1857**:274–287

[9] Slowik D, et al. Structural investigation of PsbO from plant and cyanobacterial photosystem II. Journal of Molecular Biology. 2011;**407**:125–137

[10] Umena Y, et al. Crystal structure of oxygen-evolving photosystem II at a resolution of 1.9 A. Nature. 2011; **473**:55–60

[11] Gisriel CJ, et al. High-resolution cryo-electron microscopy structure of photosystem II from the mesophilic cyanobacterium, Synechocystis sp. PCC 6803. Proceedings of the National Academy of Sciences. 2022;**119**: e2116765118

[12] Kamiya N, Shen JR. Crystal structure of oxygen-evolving photosystem II from Thermosynechococcus vulcanus at 3.7-A resolution. Proceedings of the National Academy of Sciences. 2003;**100**(1):98–103

[13] Peloquin JM, Britt RD. EPR/ ENDOR characterization of the physical and electronic structure of the OEC Mn cluster. Biochimica Et Biophysica Acta (BBA) - Bioenergetics. 2001;**1503** (1-2):96–111

[14] Yano J, et al. Structure function studies of Photosystem II using X-Ray free electron lasers. Annual Review of Biophysics. 2024;**53**:343–365

[15] Yano J, et al. X-ray damage to the Mn4Ca complex in single crystals of photosystem II: A case study for

metalloprotein crystallography. Proceedings of the National Academy of Sciences. 2005;**102**:12047–12052

[16] Lubitz W, et al. Water oxidation in photosystem II. Photosynthesis Research. 2019;**142**:105–125

[17] Bao H, Burnap RL. Photoactivation: The light-driven assembly of the water oxidation complex of Photosystem II. Frontiers in Plant Science. 2016;7:578

[18] Rivalta I, et al. Structural-functional role of chloride in photosystem II. Biochemistry. 2011;**50**:6312–6315

[19] Müh F, Zouni A. The nonheme iron in photosystem II. Photosynthesis Research. 2013;**116**(2-3):295–314

[20] Shevela D, et al. Photosystem II and the unique role of bicarbonate: A historical perspective. Biochimica Et Biophysica Acta (BBA) – Bioenergetics. 2012;**1817**:1134–1151

[21] Govindjee JJVR. Bicarbonate effects on the electron flow in isolated broken chloroplasts. Biochimica Et Biophysica Acta (BBA) - Reviews on Bioenergetics. 1978;**505**(2):183–213

[22] Gates C, et al. Photoassembly of the CaMn4O5 Catalytic core and its inorganic mutants in Photosystem II. Encyclopedia of Inorganic and Bioinorganic Chemistry. 2017;1–15

[23] Dasgupta J, et al. Photoassembly of the water-oxidizing complex in Photosystem II. Coordination Chemistry Reviews. 2008;**252**:347–360

[24] Debus RJ. Protein ligation of the photosynthetic oxygen-evolving center. Coordination Chemistry Reviews. 2008;**252**(3-4):244–258

[25] Stich TA, et al. Ligation of D1-His332 and D1-Asp170 to the manganese cluster of photosystem II from Synechocystis assessed by multifrequency pulse EPR spectroscopy. Biochemistry. 2011;**50**:7390–7404

[26] Suga M, et al. Native structure of photosystem II at 1.95 A resolution viewed by femtosecond X-ray pulses. Nature. 2015;**517**:99–103

[27] Askerka M, et al. Analysis of the radiation-damage-free X-ray structure of photosystem II in light of EXAFS and QM/MM data. Biochemistry. 2015;**54**:1713–1716

[28] Yano J, Yachandra V. Mn4 Ca cluster in Photosynthesis: where and how water is oxidized to dioxygen. Chemical Reviews. 2014;**114**(8): 4175–4205

[29] Haumann M, et al. Structural and oxidation state changes of the photosystem II manganese complex in four transitions of the water oxidation cycle (S0 –> S1, S1 –> S2, S2 –> S3, and S3,4 –> S0) characterized by X-ray absorption spectroscopy at 20 K and room temperature. Biochemistry. 2005;**44**:1894–1908

[30] Luber S, et al. S1-state model of the O2-evolving complex of photosystem II. Biochemistry. 2011;**50**:6308–6311

[31] Siegbahn PE. A structure-consistent mechanism for dioxygen formation in photosystem II. Chemistry – A European Journal. 2008;**14**(27):8290–8302

[32] Vinyard DJ, Brudvig GW. Progress toward a molecular mechanism of water oxidation in Photosystem II. Annual Review of Physical Chemistry. 2017;**68**:101–116

[33] Cox N, Messinger J. Reflections on substrate water and dioxygen formation. Biochimica Et Biophysica Acta (BBA) - Bioenergetics. 2013;**1827** (8-9):1020–1030

[34] Dau H, Haumann M. Eight steps preceding O-O bond formation in oxygenic photosynthesis–a basic reaction cycle of the Photosystem II manganese complex. Biochimica Et Biophysica Acta (BBA) - Bioenergetics. 2007;**1767**(6):472–483

[35] Kok B, et al. Cooperation of charges in photosynthetic O2 evolution–I. A linear four step mechanism. Photochemistry and Photobiology. 1970;**11**:457–475

[36] Ishikita H, Saito K. Current perspectives on proton and electron transfer pathways in photosystem II water oxidation. Plant and Cell Physiology. 2025;pcaf107. DOI: 10.1093/pcp/pcaf107

[37] Rappaport F, Lavergne J. Proton release during successive oxidation steps of the photosynthetic water oxidation process: Stoichiometries and pH dependence. Biochemistry. 1991;**30** (41):10004–10012

[38] Haumann M, et al. Photosynthetic O2 formation tracked by time-resolved x-ray experiments. Science. 2005;**310**:1019–1021

[39] Depaula JC, et al. Studies of the Manganese site of Photosystem-II by electron-spin-resonance spectroscopy. Journal of the Chemical Society, Faraday Transactions 1: Physical Chemistry in Condensed Phases. 1987;**1**:3635–3651

[40] Vinyard DJ, et al. Energetics of the S2 state spin isomers of the oxygen-evolving complex of Photosystem II. The Journal of Physical Chemistry B. 2017;**121**:1020–1025

[41] Pantazis DA, et al. Two interconvertible structures that explain the spectroscopic properties of the oxygen-evolving complex of Photosystem II in the S2 state. Angewandte Chemie International Edition. 2012;**51**:9935–9940

[42] Suga M, et al. An oxyl/oxo mechanism for oxygen-oxygen coupling in PSII revealed by an x-ray free-electron laser. Science. 2019;**366**:334–+

[43] Kern J, et al. Structures of the intermediates of Kok's photosynthetic water oxidation clock. Nature. 2018;**563**:421–425

[44] Pushkar Y, et al. Spectroscopic and computational analysis of Mn Ca cluster transformations in the oxygen evolving complex of Photosystem II. Biophysical Journal. 2020;**118**: 132a–132a

[45] Suga M, et al. Light-induced structural changes and the site of O=O bond formation in PSII caught by XFEL. Nature. 2017;**543**:131–135

[46] Pantazis DA. The S state of the oxygen-evolving complex: Overview of spectroscopy and XFEL crystallography with a critical evaluation of early-onset models for O-O bond formation. Inorganics. 2019;**7**(4)

[47] Ibrahim M, et al. Untangling the sequence of events during the S2 → S3 transition in photosystem II and implications for the water oxidation mechanism. Proceedings of the National Academy of Sciences. 2020;**117**:12624–12635

[48] Wang J, et al. Do crystallographic XFEL data support binding of a water molecule to the oxygen-evolving complex of photosystem II exposed to two flashes of light? Proceedings of the National Academy of Sciences. 2021;**118**(24):1–2

[49] Young ID, et al. Structure of photosystem II and substrate binding at room temperature. Nature. 2016;**540**:453–+

[50] Siegbahn PE. Water oxidation mechanism in photosystem II, including oxidations, proton release pathways, O-O bond formation and O2 release. Biochimica Et Biophysica Acta (BBA) - Bioenergetics. 2013;**1827** (8-9):1003–1019

[51] Kuroda H, et al. Proton transfer pathway from the oxygen-evolving complex in photosystem II substantiated by extensive mutagenesis. Biochimica Et Biophysica Acta (BBA) - Bioenergetics. 2021;**1862**(1):148329

[52] Hwang HJ, et al. Mutation of arginine 357 of the CP43 protein of photosystem II severely impairs the catalytic S state cycle of the H2O oxidation complex. Biochemistry. 2007;**46**:11987–11997

[53] Gates C, et al. Why did nature choose Manganese over Cobalt to make oxygen Photosynthetically on the earth? The Journal of Physical Chemistry B. 2022;**126**:3257–3268

[54] Capone M, et al. Reorganization of substrate waters between the closed and open cubane conformers during the S2 to S3 transition in the oxygen evolving complex. Biochemistry. 2015;**54**:6439–6442

[55] Cox N, et al. Biological water oxidation. Accounts of Chemical Research. 2013;**46**:1588–1596

[56] Najafpour MM, et al. Manganese compounds as water-oxidizing catalysts: From the natural water-oxidizing complex to nanosized manganese oxide structures. Chemical Reviews. 2016;**116**:2886–2936

[57] Cox N, et al. Electronic structure of the oxygen-evolving complex in photosystem II prior to OO bond formation. Science. 2014;**345**:804–808

[58] Noguchi T. FTIR detection of water reactions in the oxygen-evolving centre of photosystem II. Philosophical Transactions of the Royal Society B: Biological Sciences. 2008;**363** (1494):1189–1195

[59] Boussac A, Rutherford AW. Nature of the inhibition of the oxygen-evolving enzyme of photosystem II induced by sodium chloride washing and reversed by the addition of calcium (2+) or strontium (2+). Biochemistry. 1988;**27** (9):3476–3483

[60] Koua FHM, et al. Structure of Sr-substituted photosystem II at 2.1 Å resolution and its implications in the mechanism of water oxidation. Proceedings of the National Academy of Sciences. 2013;**110**: 3889–3894

[61] Gates C, et al. The strontium inorganic mutant of the water oxidizing center (CaMn4O5) of PSII improves WOC efficiency but slows electron flux through the terminal acceptors. Biochimica Et Biophysica Acta (BBA) – Bioenergetics. 2016;**1857**:1550–1560

[62] Bury G, Pushkar Y. Insights from Ca2+→ Sr2+ substitution on the mechanism of OO bond formation in photosystem II. Photosynthesis Research. 2024;**162**(2):331–351

[63] Faller P, et al. Mechanism of Cd2+ toxicity: Cd2+ inhibits photoactivation of Photosystem II by competitive binding to the essential Ca2+ site. Biochimica Et Biophysica Acta (BBA) – Bioenergetics. 2005;**1706**:158–164

[64] Siegbahn PE. Water oxidation energy diagrams for photosystem II for different protonation states, and the effect of removing calcium. Physical Chemistry Chemical Physics. 2014;**16** (24):11893–11900

[65] Zhang M, et al. Structural insights into the light-driven auto-assembly process of the water-oxidizing Mn4CaO5-cluster in photosystem II. eLife. 2017;**6**:e26933

[66] Vinyard DJ, et al. Photosystem II oxygen-evolving complex photoassembly displays an inverse H/D solvent isotope effect under chloride-limiting conditions. Proceedings of the National Academy of Sciences. 2019;**116**:18917–18922

[67] Murray JW, Barber J. Identification of a calcium-binding site in the PsbO protein of photosystem II. Biochemistry. 2006;**45**(13):4128–4130

[68] Boussac A, et al. Probing the role of chloride in Photosystem II from Thermosynechococcus elongatus by exchanging chloride for iodide. Biochimica Et Biophysica Acta (BBA) – Bioenergetics. 2012;**1817**:802–810

[69] Kawakami K, et al. Location of chloride and its possible functions in oxygen-evolving photosystem II revealed by X-ray crystallography. Proceedings of the National Academy of Sciences. 2009;**106**:8567–8572

[70] Murray JW, et al. X-ray crystallography identifies two chloride binding sites in the oxygen evolving centre of Photosystem II. Energy & Environmental Science. 2008;**1**:161–166

[71] Sakashita N, et al. Origins of water molecules in the Photosystem II crystal structure. Biochemistry. 2017;**56**:3049–3057

[72] Ghosh I, et al. Insights into Proton-Transfer pathways during water oxidation in Photosystem II. The Journal of Physical Chemistry B. 2019;**123**:8195–8202

[73] Ishida N, et al. Biosynthetic exchange of bromide for chloride and strontium for calcium in the photosystem II oxygen-evolving enzymes. Journal of Biological Chemistry. 2008;**283**:13330–13340

[74] Okamoto Y, et al. Proton and water transfer pathways in the S2 → S3 transition of the water-oxidizing complex in Photosystem II: Time-Resolved infrared analysis of the effects of D1-N298A mutation and NO3 – substitution. The Journal of Physical Chemistry B. 2021;**125**: 6864–6873

[75] Guskov A, et al. Cyanobacterial photosystem II at 2.9-A resolution and the role of quinones, lipids, channels and chloride. Nature Structural & Molecular Biology. 2009;**16**:334–342

[76] Pokhrel R, et al. Mutation of lysine 317 in the D2 subunit of photosystem II alters chloride binding and proton transport. Biochemistry. 2013;**52**: 4758–4773

[77] Raven JA. Chloride involvement in the synthesis, functioning and repair of the photosynthetic apparatus in vivo. New Phytologist. 2020;**227**(2):334–342

[78] Amin M, et al. Effect of chloride depletion on the magnetic properties and the redox leveling of the oxygen-evolving complex in Photosystem II. The Journal of Physical Chemistry B. 2016;**120**:4243–4248

[79] Jajoo A, et al. EPR characteristics of chloride-depleted photosystem II membranes in the presence of other anions. Photochemical & Photobiological Sciences. 2005;**4**:459–462

[80] Wincencjusz H, et al. The photosynthetic oxygen evolving complex requires chloride for its redox state S2–>S3 and S3–>S0 transitions but not for S0–>S1 or S1–>S2 transitions. Biochemistry. 1997;**36**:3663–3670

[81] van Vliet P, Rutherford AW. Properties of the chloride-depleted oxygen-evolving complex of photosystem II studied by electron paramagnetic resonance. Biochemistry. 1996;**35**(6):1829–1839

[82] Ono T, et al. Modification of the properties of S2 state in photosynthetic O2-evolving center by replacement of chloride with other anions. Archives of Biochemistry and Biophysics. 1987;**256**:618–624

[83] Homann PH. Structural effects of Cl - and other anions on the water oxidizing complex of chloroplast Photosystem II. Plant Physiology. 1988;**88**(1):194–199

[84] Gates C, et al. How chloride functions to enable proton conduction in photosynthetic water oxidation: Time-resolved kinetics of intermediates (S-states) in vivo and bromide substitution. Biochimica Et Biophysica Acta (BBA) - Bioenergetics. 2023;**1864**(4):148998

[85] Mandal M, et al. Requirement of chloride for the downhill electron transfer pathway from the water-splitting center in natural photosynthesis. The Journal of Physical Chemistry B. 2022;**126**:123–131

[86] Suzuki H, et al. Functional roles of D2-Lys317 and the interacting chloride ion in the water oxidation reaction of photosystem II as revealed by Fourier transform infrared analysis. Biochemistry. 2013;**52**:4748–4757

[87] Pokhrel R, et al. Probing the effect of mutations of asparagine 181 in the D1 subunit of photosystem II. Biochemistry. 2015;**54**:1663–1672

[88] Russell BP, Vinyard DJ. Chloride facilitates Mn(III) formation during photoassembly of the Photosystem II oxygen-evolving complex. Photosynthesis Research. 2022;**152**(3):283–288

[89] Miyao M, Inoue Y. Enhancement by chloride ions of photoactivation of oxygen evolution in manganese-depleted photosystem II membranes. Biochemistry. 1991;**30**(22):5379–5387

[90] Brahmachari U, et al. Chloride maintains a protonated internal water network in the photosynthetic oxygen evolving complex. The Journal of Physical Chemistry B 2017;**121**:10327–10337

[91] Hussein R, et al. Structural dynamics in the water and proton

channels of photosystem II during the S (2) to S(3) transition. Nature Communications. 2021;**12**:6531

[92] Doyle MD, et al. Water networks in Photosystem II using crystalline molecular dynamics simulations and room-temperature XFEL serial crystallography. Journal of the American Chemical Society. 2023;**145**:14621–14635

[93] Kaur D, et al. Occupancy analysis of water molecules inside channels within 25 Å Radius of the oxygen-evolving center of Photosystem II in molecular dynamics simulations. The Journal of Physical Chemistry B. 2024;**128**:2236–2248

[94] Hussein R, et al. Evolutionary diversity of proton and water channels on the oxidizing side of photosystem II and their relevance to function. Photo synthesis Research. 2023;**158**:91–107

[95] Allgower F, et al. Molecular principles of redox-coupled protonation dynamics in photosystem II. Journal of the American Chemical Society. 2022;**144**:7171–7180

[96] Stemler A, Govindjee. Bicarbonate ion as a critical factor in photosynthetic oxygen evolution. Plant Physiology. 1973;**52**(2):119–123

[97] Tikhonov K, et al. Quantification of bound bicarbonate in photosystem II. Photosynthetica. 2018;**56**:210–216

[98] Klimov VV, et al. Bicarbonate requirement for the donor side of photosystem II. FEBS Letters. 1995;**363**:251–255

[99] Van Rensen JJ, Xu C. Role of bicarbonate in photosystem II, the water-plastoquinone oxido-reductase of plant photosynthesis. Physiologia Plantarum. 1999;**105**(3):585–592

[100] Banerjee G, et al. Bicarbonate rescues damaged proton-transfer pathway in photosystem II. Biochimica Et Biophysica Acta (BBA) – Bioenergetics. 2019;**1860**:611–617

[101] Stemler A. Inhibition of Photosystem-II by formate - possible evidence for a direct role of bicarbonate in photosynthetic oxygen evolution. Biochimica Et Biophysica Acta (BBA) - Bioenergetics. 1980;**593**(1):103–112

[102] Stemler AJ. The bicarbonate effect, oxygen evolution, and the shadow of Otto Warburg. Photosynthesis Research. 2002;**73** (1-3):177–183

[103] Noguchi T, Inoue Y. Identification of Fourier transform infrared signals from the non-heme iron in photosystem II. The Journal of Biochemistry. 1995;**118**(1):9–12

[104] Saito K, et al. Mechanism of proton-coupled quinone reduction in Photosystem II. Proceedings of the National Academy of Sciences. 2013;**110**:954–959

[105] Sugo Y, Ishikita H. Proton-mediated photoprotection mechanism in photosystem II. Frontiers in Plant Science. 2022;**13**:934736

[106] Ananyev G, et al. The multiplicity of roles for (bi) carbonate in photosystem II operation in the hypercarbonate-requiring cyanobacterium Arthrospira maxima. Photosynthetica. 2018;**56**:217–228

[107] Vermaas WFJ, Rutherford AW. Electron-Paramagnetic-Res measurements on the effects of

bicarbonate and triazine resistance on the acceptor side of photosystem-II. FEBS Letters. 1984;**175**(2):243–248

[108] Xiong J, et al. Modeling of the D1/D2 proteins and cofactors of the photosystem II reaction center: Implications for herbicide and bicarbonate binding. Protein Science. 1996;**5**:2054–2073

[109] Hussein R, et al. Cryo-electron microscopy reveals hydrogen positions and water networks in photosystem II. Science. 2024;**384**:1349–1355

[110] Kato K, et al. High-resolution cryo-EM structure of photosystem II reveals damage from high-dose electron beams. Communications Biology. 2021;**4**:382

[111] Xiong J, et al. Loss of inhibition by formate in newly constructed photosystem II D1 mutants, D1-R257E and D1-R257M, of Chlamydomonas reinhardtii. Biochimica Et Biophysica Acta (BBA) – Bioenergetics. 1998;**1365**:473–491

[112] Takahashi R, et al. Structural coupling of a tyrosine side chain with the non-heme iron center in photosystem II as revealed by light-induced Fourier transform infrared difference spectroscopy. Biochemistry. 2009;**48**:8994–9001

[113] Gisriel CJ, et al. Structure of a monomeric photosystem II core complex from a cyanobacterium acclimated to far-red light reveals the functions of chlorophylls d and f. Journal of Biological Chemistry. 2022;**298**(1):101424

[114] Zabret J, et al. Structural insights into photosystem II assembly. Nature Plants. 2021;**7**:524–538

[115] Wydrzynski T, Govindjee. A new site of bicarbonate effect in photosystem II of photosynthesis: Evidence from chlorophyll fluorescence transients in spinach chloroplasts. Biochimica Et Biophysica Acta (BBA) - Bioenergetics. 1975;**387**(2):403–408

[116] Brinkert K, et al. Bicarbonate-induced redox tuning in Photosystem II for regulation and protection. Proceedings of the National Academy of Sciences. 2016;**113**:12144–12149

[117] Fantuzzi A, et al. Bicarbonate-controlled reduction of oxygen by the QA semiquinone in Photosystem II in membranes. Proceedings of the National Academy of Sciences. 2022;**119**(6)

[118] Forsman JA, Eaton-Rye JJ. The D1: Ser268 residue of Photosystem II contributes to an alternative pathway for Q B protonation in the absence of bound bicarbonate. FEBS Letters. 2020;**594**(18):2953–2964

[119] Jursinic P, et al. A major site of bicarbonate effect in system II reaction. Evidence from ESR signal IIvf, fast fluorescence yield changes and delayed light emission. Biochimica Et Biophysica Acta (BBA) – Bioenergetics. 1976;**440**:322–330

[120] Govindjee, et al. Inhibition of the reoxidation of the secondary electron acceptor of photosystem II by bicarbonate depletion. Biochimica Et Biophysica Acta (BBA) – Bioenergetics. 1976;**449**:602–605

[121] Khanna R, et al. Evidence for a close spatial location of the binding-sites for CO2 and for photosystem-II inhibitors. Biochimica Et Biophysica Acta (BBA) – Bioenergetics. 1981;**634**:105–116

[122] Eatonrye JJ, Govindjee. Electron-Transfer through the quinone acceptor complex of photosystem-II in bicarbonate-depleted spinach thylakoid membranes as a function of actinic flash number and frequency. Biochimica Et Biophysica Acta (BBA) - Bioenergetics. 1988;**935**(3):237–247

[123] Blubaugh DJ, Govindjee. The molecular mechanism of the bicarbonate effect at the plastoquinone reductase site of photosynthesis. Photosynthesis Research. 1988;**19**(1-2):85–128

[124] Kato Y, Noguchi T. Redox properties and regulatory mechanism of the iron-quinone electron acceptor in photosystem II as revealed by FTIR spectroelectrochemistry. Photosynthesis Research. 2022;**152**(2):135–151

[125] Sugo Y, et al. Conformational changes and H-Bond rearrangements during quinone release in Photosystem II. Biochemistry. 2022;**61**:1836–1843

[126] Castillo L, et al. Depletion reveals role of bicarbonate in the photosynthetic electron transport chain of Limnospira maxima. Frontiers in Plant Science. 2025;**16**:1584909

[127] Kato Y, et al. Redox potential of the terminal quinone electron acceptor Q B in photosystem II reveals the mechanism of electron transfer regulation. Proceedings of the National Academy of Sciences. 2016;**113**:620–625

[128] Ananyev G, et al. Mutagenesis of CP43-arginine-357 to serine reveals new evidence for (Bi)carbonate functioning in the water oxidizing complex of Photosystem II. Photochemical & Photobiological Sciences. 2005;**4**:991–998

[129] Vinyard DJ, Govindjee G. Bicarbonate is a key regulator but not a substrate for O(2). Photosynthesis Research. 2024;**162**:93–99

[130] Koroidov S, et al. Mobile hydrogen carbonate acts as proton acceptor in photosynthetic water oxidation. Proceedings of the National Academy of Sciences. 2014;**111**:6299–6304

[131] Baranov SV, et al. Bicarbonate accelerates assembly of the inorganic core of the water-oxidizing complex in manganese-depleted photosystem II: A proposed biogeochemical role for atmospheric carbon dioxide in oxygenic photosynthesis. Biochemistry. 2000;**39**:6060–6065

[132] Baranov SV, et al. Bicarbonate is a cofactor for assembly of the manganese cluster of the photosynthetic water oxidizing complex. Kinetics of reconstitution of O evolution by photoactivation. Biochemistry. 2004;**43**:2070–2079

[133] Klimov VV, et al. Bicarbonate may Be required for ligation of manganese in the oxygen-evolving complex of photosystem II. Biochemistry. 1997;**36**:16277–16281

[134] Dasgupta J, et al. Bicarbonate coordinates to Mn during photo-assembly of the catalytic MnCa core of photosynthetic water oxidation: EPR characterization. Applied Magnetic Resonance. 2010;**37**:137–150

[135] Diner BA, Petrouleas V. Q400, the non-heme iron of the photosystem II iron-quinone complex. A spectroscopic probe of quinone and inhibitor binding to the reaction center. Biochimica et Biophysica Acta (BBA)-Reviews on Bioenergetics. 1987;**895**(2):107–125

[136] Shinkarev VP. Flash-induced oxygen evolution in photosynthesis: Simple solution for the extended S-state model that includes misses, double-hits, inactivation, and backward-transitions. Biophysical Journal. 2005;**88**(1):412–421

[137] Shevela D, et al. Interactions of photosystem II with bicarbonate, formate and acetate. Photosynthesis Research. 2007;**94**:247–264

[138] Bowes JM, et al. A high potential acceptor for photosystem II. Biochimica Et Biophysica Acta (BBA) – Bioenergetics. 1979;**547**:320–335

[139] Ishikita H, Knapp E-W. Oxidation of the non-heme iron complex in photosystem II. Biochemistry. 2005;**44**(45):14772–14783

[140] Pospíšil P. The role of metals in production and scavenging of reactive oxygen species in photosystem II. Plant and Cell Physiology. 2014;**55**(7):1224–1232

[141] Bernarding J, et al. Kinetic studies on the stabilization of the primary radical pair P680+ Pheo-in different photosystem II preparations from higher plants. Photochemistry and Photobiology. 1994;**59**:566–573

[142] MacMillan F, et al. EPR and ENDOR investigation of the primary electron acceptor radical anion QA. bul.-in iron-depleted photosystem II membrane fragments. Biochemistry. 1995;**34**:8144–8156

[143] Astashkin AV, et al. A comparative electron spin echo envelope modulation study of the primary electron acceptor quinone in Zn-substituted and cyanide-treated preparations of photosystem II. The Journal of Chemical Physics. 1998;**108**:10143–10151

[144] Boussac A, et al. Probing the quinone binding site of Photosystem II from thermosynechococcus elongatus containing either PsbA1 or PsbA3 as the D1 protein through the binding characteristics of herbicides. Biochimica Et Biophysica Acta (BBA) – Bioenergetics. 2011;**1807**:119–129

[145] Kato Y, Noguchi T. Long-range interaction between the Mn4CaO5 cluster and the non-heme iron center in photosystem II as revealed by FTIR spectroelectrochemistry. Biochemistry. 2014;**53**(30):4914–4923

Chapter 4

Genetic Modifications in Cyanobacteria: Current Status and Future Prospects

Kritika Tripathi and Yogesh Mishra

Abstract

Cyanobacteria are powerful model systems for fundamental and applied research, ranging from photosynthesis and nitrogen fixation to biofuel and biomolecule production. Over the past five decades, the field of cyanobacterial genetics has transformed from basic mutagenesis and plasmid transfer methods into a sophisticated synthetic biology discipline supported by advanced molecular and computational tools. Early developments, such as the discovery of natural competence, conjugative DNA transfer, and shuttle vector construction, established the foundation for stable transformation systems in diverse strains. The subsequent identification of strong native and heterologous promoters, synthetic ribosome binding sites, and optimized selection markers further refined the control of gene expression. Genome sequencing of cyanobacterial strains enabled vector design and genome-scale metabolic engineering. More recently, the advent of markerless selection systems and CRISPR/Cas-based genome editing tools has revolutionized the precision and efficiency of genetic manipulation in cyanobacteria. This chapter provides an integrated overview of the history, evolution, and applications of genetic tools in cyanobacteria. It uniquely covers progress from classical recombination to CRISPR technologies and the emerging future consisting of AI-driven gene design, multiomics integration, and minimal chassis development, while highlighting challenges in achieving standardized, stable, and scalable cyanobacterial strains for sustainable biotechnology.

Keywords: CRISPR/Cas, cyanobacteria, genetic manipulation, genetic tools, markers, plasmids, promoters, ribosome-binding sites, riboswitches, shuttle vectors

1. Introduction

Genetic modification can be defined as the deliberate alteration of a cell's or an organism's DNA content in order to produce a desired trait for applications in the fields of medicine, research, and agriculture [1]. Humans have been performing genetic manipulations for thousands of years by selectively breeding organisms with desired traits (e.g., the domestication of dogs, cabbage, and chicken) [2]. In modern times, it relies on sophisticated molecular techniques to add, overexpress, or silence gene sequences in an organism to achieve a targeted phenotype or product in a sustainable way [1].

In this context, cyanobacteria are promising candidates for sustainable biotechnology because they efficiently use sunlight and CO_2 to produce diverse metabolites, require simple nutrients, and are genetically easy to modify [3]. Genetic modification in cyanobacteria offers a versatile platform for the sustainable production of biomass and renewable biofuels, including ethanol, butanol, biodiesel, and hydrogen [4–7]. Beyond energy applications, engineered cyanobacteria hold significant promise in environmental and industrial biotechnology. They can be used for bioremediation of heavy metals and other pollutants, biofertilization to enhance soil nutrient content, and aquaculture to improve water quality and support feed production [8]. Additionally, they can serve as biofactories for the synthesis of biologically active molecules and high-value compounds, such as vitamins, pigments, antioxidants, and pharmaceuticals. This capability helps bridge the gap between sustainable production and commercial application [9–12].

From the discovery of the DNA double helix by Watson and Crick in 1953 [13] to the recently discovered and emerging technologies like CRISPR-Cas [14], the field of genetic engineering has seen a series of transformative discoveries that have made genetic manipulation easy and quick. So far, genetic engineering has been implemented in a wide range of organisms, ranging from bacteria to humans, but the availability and advancement of genetic engineering tools are not the same across different species [15]. Simple, well-characterized organisms like bacteria and yeast, with less complex genomes, have robust and well-established genome editing tools, while complex organisms like cyanobacteria continue to present challenges. The gene editing techniques discovered for heterotrophic bacteria, such as *Escherichia coli,* cannot be directly implemented on cyanobacteria due to differences in the cellular organization and physiology of these organisms [15]. The differences that make cyanobacterial systems challenging include the presence of thylakoid membranes, mucilaginous sheaths [16], specialized cells like heterocysts and akinetes, along with multiple genome copies per cell (polyploidy), which complicate genetic segregation and stable mutant generation [17–19]. In addition, many cyanobacteria possess strong restriction–modification systems that degrade foreign DNA, making transformation inefficient [20–22]. They also possess a very efficient recombination machinery [23], which not only facilitates the integration of exogenous DNA by homologous recombination but can also result in the unintended integration of replicative vectors, making mutant behavior unpredictable. Finally, the presence of circadian rhythms [24] can lead to time-dependent behavior in engineered genetic networks.

Furthermore, standardized genetic parts, such as promoters, ribosome-binding sites (RBSs), and terminators, often behave unpredictably in cyanobacteria because their transcriptional and translational machinery, light-regulated gene expression, and regulatory networks differ from those of model bacteria [25, 26]. Therefore, these regulatory elements must be characterized and optimized within cyanobacterial systems for consistent and predictable gene expression outcomes.

Nevertheless, the rapid expansion of genome sequencing and annotation data for diverse cyanobacterial species over the past few decades has greatly facilitated the development of molecular tools for their genetic manipulation. Transfer of exogenous DNA into unicellular or filamentous strains has been made easy using methods like natural transformation, electroporation, and conjugation with self-replicating plasmids or integrative vectors. A large number of positive and negative selection markers, along with reporter constructs, have also been established for promoter activity measurements and protein localization studies. Moreover, extensive knowledge of restriction–modification systems in several cyanobacterial strains has been acquired to improve transformation efficiency [27]. The advancement of genetic tools, including metabolic engineering, synthetic biology, and CRISPR-based genome editing, has significantly expanded the potential of cyanobacteria, allowing precise manipulation of metabolic pathways to enhance productivity, stress tolerance, and environmental adaptability [28]. In this chapter, we are going to summarize the history, evolution, and existing knowledge of genetic manipulation tools in cyanobacteria, along with their applications and future prospects.

2. History and evolution of genetic manipulation tools in cyanobacteria

The discovery of the DNA double helix [13], followed by the identification of ligases, restriction enzymes, and plasmid cloning systems [29–32], laid the foundation of recombinant DNA technology. This was later revolutionized by the establishment of the polymerase chain reaction in the 1980s [33]. While genetic engineering was advancing across diverse systems, cyanobacterial genetic manipulation tools also gradually evolved from the 1970s through the 1990s, building the foundation for modern synthetic biology applications. Early studies include the first experimental report demonstrating natural genetic transformation (i.e., uptake and incorporation of exogenous DNA) in a cyanobacterium (*Anacystis nidulans*, now known as *Synechococcus elongatus* PCC 6301), among others [34]. Another study revealed that many cyanobacterial strains carry circular plasmid DNAs (CCC DNAs) with inter-strain homologies and potential for horizontal transfer, suggesting the possibility for the development of plasmid-based cloning tools [35]. Simultaneously, classical genetic selections and mutant screens, like amino acid antimetabolites for regulatory mutants, were being established, providing early genetic systems for cyanobacteria [36].

In the 1980s and early 1990s, cyanobacterial genetic manipulation research focused on establishing DNA transformation methods while developing shuttle vectors, selectable markers, and transposon systems for mutagenesis. The first successful attempts to introduce foreign DNA into cyanobacteria through natural competence, conjugation, and electroporation were reported [37–39, 40]. These classical approaches were complemented by transposon mutagenesis with transposon systems such as Tn5 and Tn9, which allowed random gene disruption for functional analysis [41–43]. Site-directed mutagenesis for precise gene substitution by homologous recombination was also established [44, 45]. Complementation assays were developed to confirm gene function by recovering mutant phenotypes with wild-type copies [46, 47].

A major breakthrough came with the introduction of conjugation between *Escherichia coli* and cyanobacteria, mediated by broad-host-range IncP plasmids such as RP4. This enabled efficient gene transfer across diverse unicellular and filamentous cyanobacteria [37, 48, 49]. Electroporation soon followed as a simpler and more controlled alternative for introducing DNA [39, 50]. However, restriction–modification systems in cyanobacteria often degraded unmethylated foreign DNA, reducing transformation efficiency [38, 51]. To address this, shuttle vectors were developed with dual replication origins for *E. coli* and cyanobacteria [51–53]. Among these, IncQ-derived plasmids such as RSF1010 exhibited wide host range and stability, functioning efficiently in several species [54–56].

Around this time, selectable markers, such as antibiotic resistance genes conferring tolerance to neomycin, erythromycin, spectinomycin, chloramphenicol, and kanamycin, were widely used. Hybrid cassettes combining native cyanobacterial promoters with bacterial resistance genes improved expression and selection efficiency in both unicellular and heterocystous strains [57, 58].

The early 1990s marked a milestone with promoter–reporter fusion systems that allowed in vivo monitoring of gene regulation. Reporters such as lacZ (β-galactosidase) and luxAB (luciferase) were used to quantify promoter activity [59, 60]. Strong constitutive promoters such as P*psbA2*, P*cpcB*, and phage-derived promoters like λPR and λPL were identified for high-level gene expression [48, 61, 62]. Regulatable promoters, including P*tac* (LacI-controlled) and phage T7 promoters, which were an early attempt at orthogonal expression, were also reported, but their use was limited due to challenges like inducer cost and heat sensitivity [63, 64]. Nevertheless, these systems paved the way for modern synthetic promoters.

An important turning point came with the mapping and sequencing of cyanobacterial genomes in the early 1990s [65]. By 1996, the complete genome sequence of *Synechocystis* sp. PCC 6803 was published, marking the first photosynthetic organism to be fully sequenced and revolutionizing the field by providing genetic blueprints [66]. Shortly after this, the CyanoBase database was established [67] to add and annotate cyanobacterial genomic information. Genome sequencing of different cyanobacterial strains enabled the identification of neutral integration sites, native promoters, and restriction–modification systems, accelerating the design of vectors, promoters, and other transcriptional and translational entities, as discussed in the following sections.

Since the 2000s, the cyanobacterial genetic toolkit has expanded rapidly with optimized shuttle vectors, modular cloning systems (BioBrick, CYANO-VECTOR, CyanoGate, etc.) [68–70], engineered promoters and RBS libraries [71], inducible systems adapted to cyanobacterial physiology, and the onset of CRISPR-based editing. CRISPR–Cas technology was adapted in cyanobacteria in 2014, first demonstrated in *Synechococcus elongatus* PCC 7942 [72]. Subsequent advancements enabled efficient, markerless genome editing [73] and transcriptional regulation through CRISPR interference (CRISPRi) [74, 75], followed by Cas12a-based and multiplexed CRISPRi/a systems that now support genome-scale manipulation across species [76]. This period also witnessed a surge in the application of genetically engineered cyanobacteria for industrial and biotechnological purposes, such as biofuel, bioplastic, and high-value compound production. Additionally, they served as model systems to elucidate fundamental biological processes, including gene function, photosynthetic regulation, and transcriptional control [6, 77–79].

In summary, cyanobacterial genetic manipulation evolved from classical microbiological genetics and recombinant-DNA techniques in the 1970s, through shuttle vectors and reporter systems in the 1990s–2000s, to the modern synthetic biology and CRISPR era of today, as shown in **Figure 1**.

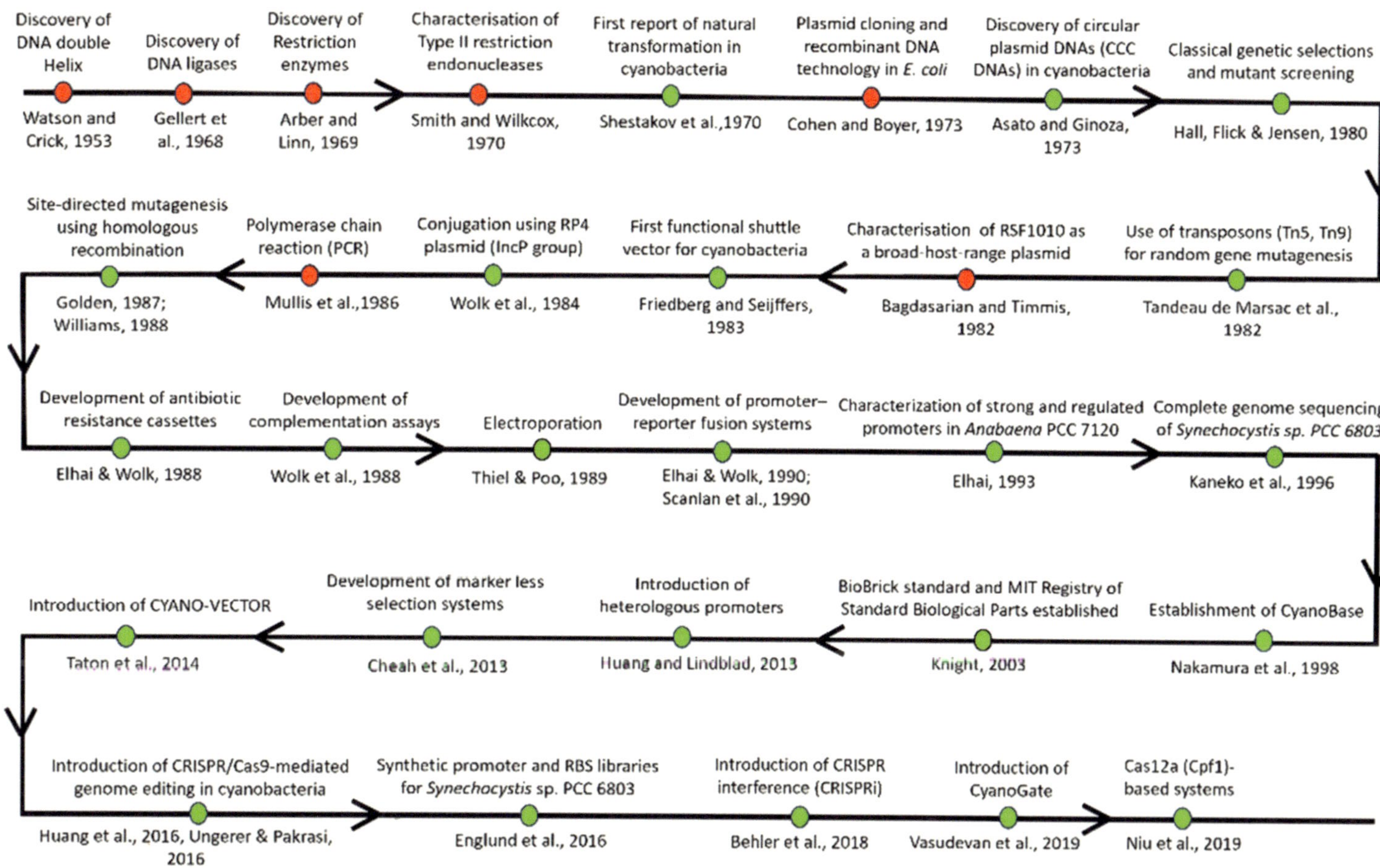

Figure 1.
Chronological overview of key events in the progress of cyanobacterial genetic manipulation (1970– 2020). The red points show some of the general milestones that laid the foundation of genetic engineering [13, 29–33, 54] while the green points are key events specific to cyanobacteria [26, 34–39, 41, 44–46, 52, 59–61, 66–76].

3. Genetic tools for cyanobacteria

Effective genetic modification of cyanobacteria requires some basic tools, called genetic tools, that help in controlling gene expressions and post-translational regulations. Many genetic tools, like neutral sites, integrative or shuttle vectors, promoters, ribosome binding site (RBS) sequences, riboswitches, markers, CRISPR/Cas systems, etc., have been developed for both basic biological research and strain engineering in cyanobacteria. These tools have been successfully used in studies involving gene transfer, gene expression, gene control, and metabolic modification [80–82]. Currently, the development and application of genetic tools in cyanobacteria are largely lagging behind the toolboxes widely available for *E. coli, Bacillus subtilis,* and *S. cerevisiae* due to the earlier-mentioned challenges [15]. In this section, we discuss some of the available genetic tools in cyanobacteria, their current status, and limitations.

3.1 Neutral sites

Heterologous gene insertion in cyanobacteria mainly relies on transformation and homologous recombination, for which researchers often use neutral chromosomal sites that do not affect host physiology (e.g., *slr0168* [81], *psbA1* in *Synechocystis* [83], NSI/NSII in *S. elongatus 7942* [84], *and glpK* in *Synechococcus 7002* [85]). *A detailed list of neutral sites in Synechocystis and other cyanobacterial strains can be found in the literature* [86, 87]. However, the number of known neutral sites is limited, and transformation efficiency in such cases depends on natural competence, which makes this method restricted to established model cyanobacteria [88]. Furthermore, the polyploidy of species like *Synechocystis* makes complete segregation laborious and time-consuming [19], limiting the flexibility and speed of chromosomal integration for complex genetic engineering.

3.2 Replicative plasmids and shuttle vector systems in cyanobacteria

Replicative plasmids are extrachromosomal DNA elements that replicate autonomously, either through host-encoded proteins or their own replication machinery. Unlike chromosomal integration, plasmid-based systems avoid segregation issues in polyploid cyanobacteria and thus enable faster genetic manipulation [87]. Many native plasmids confer traits such as antibiotic resistance or stress tolerance [89] and vary in size, replication origin, and copy number, ranging from low (2–10 copies) to high (>20–100 copies) per cell. Synthetic plasmids often include selectable markers and multiple cloning sites; however, conventional *E. coli* vectors, such as pUC [90], do not replicate in cyanobacteria, leading to the need for shuttle vectors.

Among the few broad-host-range shuttle vectors, derivatives of RSF1010 are the most extensively used and can replicate in over 30 Gram-negative bacteria, including *Synechocystis*, *Synechococcus*, and *Anabaena* [69, 91]. Several optimized RSF1010 derivatives, such as pPMQAK1, pSL1211, pFC1, pDF, pSHDY, and the pSEVAx5n series, have been developed to reduce size and improve stability. Another broad-host-range plasmid, RK2, also replicates in *Synechocystis* [92]. Since only one plasmid of the same incompatibility group can stably coexist in a host, synthetic shuttle vectors have been developed by fusing E. coli-compatible replication systems with

Endogenous/native plasmid and host strain	Shuttle vector or shuttle vector series	Reported host strains	References
pCA2.4 (*Synechocystis* sp. PCC 6803)	pCA-UC118, pSOMA series	*Synechocystis* sp. PCC 6803, *Nostoc* sp. PCC 7120	[95, 96]
pCB2.4 (*Synechocystis* sp. PCC 6803)	pCB-SC101, pSOMA series	*Synechocystis* sp. PCC 6803	[95, 96]
pCC5.2 (*Synechocystis* sp. PCC 6803)	pSCB series, pCCM1-FbFP, pYS series	*Synechocystis* sp. PCC 6803, *Synechococcus elongatus* PCC 7942, *Synechococcus* sp. PCC 7002, *Nostoc* sp. PCC 7120	[97–99]
pSYSA (*Synechocystis* sp. PCC 6803)	VIII series	*Synechocystis* sp. PCC 6803, *Synechococcus elongatus* PCC 7942	[100]
pSYSX (*Synechocystis* sp. PCC 6803)	p6031 and p6090 series	*Synechococcus elongatus* PCC 7942	[101]
pANS/pUH24 (*Synechococcus* elongatus PCC 7942)	pLS103, pSG111, pAM series	*Synechococcus elongatus* PCC 7942, *Nostoc* sp. PCC 7120	[101–104]
pAQ1 (*Synechococcus* sp. PCC 7002)	pAQE series	*Synechococcus* sp. PCC 7002	[105]
pDU1 (*Nostoc* sp. PCC 7524)	pRL series, pRL2697, pRL25C, pDUCA7, pRL489, pRL1050	*Nostoc* sp. PCC7120, *Nostoc* sp. PCC 7524, *Anabaena* sp. M-131, *Fischerella muscicola* PCC 7414 and *Chlorogloeopsis fritschii* PCC 6912, *Chroococcidiopsis* spp. str. 029, 057, and 123, *Oscillatoria* MKU 277, *Fischerella thermalis* SAG 4.84	[37, 46, 61, 64, 106–110]
pDC1 (*Nostoc* sp. MAC PCC 8009)	pSCR119/202, pSUN119/202	*Nostoc punctiforme*	[111, 112]
pGL3 (*Plectonema boryanum* sp. PCC 6306)	pPBH201	*Nostoc* sp. PCC 7120, *Plectonema boryanum* sp. PCC 6306	[113]
pFDA (*Fremyella diplosiphon*)	pJCF series	*Fremyella diplosiphon*	[114]

Table 1.
List of shuttle vectors derived from endogenous/native plasmids of cyanobacteria and their reported host strains (apart from E. coli*).*

cyanobacteria-specific replication elements. These vectors function in both unicellular species (*Synechocystis, S. elongatus* PCC 7942, *Synechococcus* PCC 7002) and filamentous strains like *Anabaena* sp. PCC 7120, which are difficult to transform through homologous recombination [87]. Nevertheless, mechanistic understanding of plasmid replication, copy number regulation, and maintenance in cyanobacteria remains incomplete [93]. A comprehensive list of cyanobacterial shuttle vectors, along with their host strains, is given in **Table 1**.

In parallel, modular cloning systems have been designed to standardize cyanobacterial genetic engineering. The CYANO-VECTOR platform [69] and

CyanoGate [70], based on Golden Gate MoClo standards, facilitate rapid assembly of multigene constructs compatible with several hosts [94]. These systems, together with the increasing availability of endogenous plasmid-derived shuttle vectors, have significantly expanded the cyanobacterial genetic toolbox, enabling modular, high-level gene expression and advanced synthetic biology applications across diverse cyanobacterial species.

3.3 Promoters

For successful heterologous gene expression in cyanobacteria, it is not only essential to introduce and maintain foreign genes but also to ensure their optimal expression for achieving a high yield of targeted products. This is achieved by promoter engineering. Promoter engineering is a very effective and frequently used tool for controlling gene expression. Promoters are specific DNA sequences that recruit RNA polymerase and initiate transcription. They can be classified based on origin – endogenous (produced from the host) or heterologous promoters (produced from another organism), and activity – constitutive (which are always active) or inducible promoters (which require an additional regulatory element) [87].
A comprehensive list of commonly used promoters, their inducers (in the case of inducible promoters), and host strains has been given in **Table 2**.

- Endogenous or native promoters

  Endogenous promoters, or native promoters such as photosystem promoters, have been extensively studied for heterologous gene expression in cyanobacteria, but most exhibit light-dependent activity and limited predictability,

Inducible promoters			
Name	**Inducer**	**Host strains**	**References**
PL03	aTc ligand	*Synechocystis* 6803	[26]
Psca6-2	IPTG	*Synechocystis* 6803	[115]
PO2	FNR under dark conditions	*Synechocystis* 6803	[116]
PompC	Envz under dark conditions	*Synechocystis* 6803	[117]
Pvan	Vanillate	*Synechococcus* 7942	[118]
PBAD	L-arabinose	*Synechococcus* 7942, *Synechocystis* 6803	[117, 119]
PrhaBAD	Rhamnose	*Synechocystis* 6803	[120]
PisiAB	Fe^{2+}	*Synechocystis* 6803, *Synechococcus elongatus* 7942, *Anabaena*, and *Leptolyngbya* BL0902	[69, 81, 121]
PphoA	Inorganic phosphorus	*Synechocystis*, *Synechococcus elongatus* sp. PCC 7942, *Anabaena*, and *Leptolyngbya* BL0902	[69]
PnrsB	Ni^{2+} or Co^{2+}	*Synechocystis*	[95, 122–126]
PpetE	Cu^{2+}	*Synechocystis*, *Anabaena*	[71, 127–133]

Inducible promoters		
Constitutive promoters		
Name	**Host strains**	**References**
PR–PS	*Synechococcus* 7942	[134]
Pcpc560	*Synechocystis* and *S. elongatus* UTEX2973	[70, 135–138]
PpsbA	*Synechocystis*, *S. elongatus* 7942, *Anabaena*, and *Synechococcus* 7002	[61, 139–141]
PpsbA1	*S. elongatus* 7942, *Synechocystis*, *Anabaena*, and *Leptolyngbya* BL0902	[4, 69, 142–144]
PpsbA2S	*Synechocystis*	[71, 139, 145–147]
PA2520	*Synechococcus* 7002	[148]
PA2579	*Synechococcus* 7002	[148]
Plac	*Synechococcus* 2973	[149]
Ptrc	*Synechococcus* 2973	[150]
BioBrick BBa_J23101	*Synechocystis*, *S. elongatus* 7942, and *Anabaena*	[26, 71, 72, 95, 128, 151]
BioBrick BBa_J23119	*Synechocystis*, *S. elongatus* 7942, *Synechococcus* 7002, *Anabaena*, and *S. elongatus* UTEX2973	[70–72, 74, 136, 152, 153]
PconII	*Synechocystis*, *S. elongatus* 7942, *Synechocystis* sp. Strain WHSyn, *Anabaena*, and *Leptolyngbya* BL0902	[69, 72, 92, 154]
Prbc	*Synechocystis*	[71, 129, 146, 155]
PrnpB	*Synechocystis*	[71, 128, 155]
Ptrc1O	*Synechocystis*	[26, 127, 128, 139, 155]

Table 2.
A list of promoters developed for genetic engineering in cyanobacteria.

making them less suitable for metabolic engineering [3, 142, 156, 157]. Most of these promoters are derived from genes related to photosynthesis, such as CO2 fixation (P*rbcL*, P*cmp*, P*sbt*), photosystem I (PSI; P*psaA*, P*psaD*), PSII (P*psbA1*, P*psbA2*), and the photosynthesis antenna protein phycocyanin (P*cpc*) [158]. Studies in *Synechococcus* PCC 7002 involving the expression of heterologous yellow fluorescent protein (YFP) showed that most native promoters yielded fluorescence similar to background wild-type fluorescence [148], indicating that native promoters require additional design to achieve the desired expression of heterologous genes [3]. Similarly, experiments conducted on *Synechococcus* PCC 7942 to increase the expression of limonene synthase for the production of limonene revealed that only one native promoter among all tested promoters produced as much limonene as the heterologous Ptrc promoter, while a heterologous pea psbA promoter achieved the highest expression [159]. This is because native promoters often contain unknown regulatory elements that can cause unpredictable gene expression and genetic instability, where strains suddenly drop in productivity or performance [160]. Such issues are further complicated by the polyploid nature of cyanobacterial chromosomes, making full segregation difficult and leading to variability in productivity. These findings emphasize the need for more tightly regulated, well-characterized heterologous promoters and genetic parts to ensure stable and predictable gene expression in cyanobacteria [3].

- Heterologous promoters

  Heterologous promoters have also been explored in cyanobacteria to overcome native regulatory constraints and avoid strain instability from repeated sequences. Common *E. coli* promoters, such as λPL, λPR, and P*Lac*, show little to no activity in *Synechocystis* PCC 6803, with minimal reporter fluorescence detected from P*lac*, PR, P*trc*, and P*tet* [155]. Modifying elements like the - 10 region to the consensus sequence (TATAAT) can improve expression in *E. coli* but not in cyanobacteria [26], because fully consensus promoters may reduce activity due to excessive contact between RNA polymerase and the promoter [161]. Promoter performance also varies between cyanobacterial strains. For example, in *S. elongatus* sp. PCC 7942, heterologous promoters P*trc*, P*LlacO1*, P*conII*, P*J23101*, and P*J23119* were tested, with P*conII* and PJ23119 giving the highest expression [72].

  Heterologous promoters behave unpredictably in cyanobacteria due to fundamental differences in their transcription machinery compared to model bacteria like *E. coli*. Cyanobacterial promoters generally have a strong - 10 element (TATAAT) but lack a well-defined - 35 consensus, suggesting reliance on sigma factors for promoter recognition [162]. Promoters in cyanobacteria are classified into three types: Type I (σ70; contains both - 10 and - 35), Type II (-10 plus activator-binding sites, e.g., glnB P2 regulated by NtcA), and Type III (lacking consensus elements; stress-responsive) [128]. Sigma factors are grouped into Group 1, primary (sigA); Group 2, nonessential/environment-responsive (sigB, sigC, sigD, sigE); and Group 3, stress-specific (sigF, sigG, sigH, sigI) families, but limited understanding of their interactions hampers prediction of promoter activity [163, 164]. Further differences in RNA polymerase structure, sigma factor specificity, and overall transcriptional regulation have been extensively described in the literature [3]. These variations help explain why heterologous promoters frequently exhibit altered or inconsistent performance in cyanobacteria compared to *E. coli*.

- Inducible promoters

  Inducible promoters are crucial tools in cyanobacterial genetic engineering, as they allow precise control over gene expression. Ideally, such promoters show minimal basal activity, respond rapidly to inducers, and exhibit tunable expression proportional to inducer concentration [93]. They are particularly valuable when the introduced genes cause a metabolic overload or produce toxic byproducts [165]. Native inducible promoters in cyanobacteria respond to various environmental or metal ion cues. Examples include the nickel- and cobalt-inducible P*nrsB* [71], the copper-responsive P*petE* and *CoaA* [127], the zinc-responsive P*smt*, the nitrate/nitrite-inducible P*nirA*, and the light-regulated PpsbA [154]. Among these, P*nrsB* is one of the most effective, providing tight regulation and strong induction upon exposure to Ni^{2+} or Co^{2+} [71]. However, many native metal-inducible systems face limitations, such as inducer toxicity at higher concentrations and interference with cellular metabolism [71, 127, 155, 165, 166].

Several heterologous induction systems from *E. coli*, like the IPTG-inducible P*trc* [83, 167], P*tac* [61], and P*LlacO1* [6], have also been tested in different cyanobacteria, with reports of both strong and poor expression depending on the host strain and construct design. To expand the toolbox, new heterologous systems, like the anhydrotetracycline (aTc)-inducible P*L03* promoter [26] and sugar-inducible systems such as P*BAD* (L-arabinose-inducible) and P*rhaBAD* (rhamnose-inducible) [120], showing stable and tunable expression with minimal leakiness across multiple cyanobacterial species, have been successfully established.

In summary, both native and synthetic inducible promoters have been explored in cyanobacteria. While native metal- and light-responsive promoters provide host compatibility, heterologous systems offer orthogonal regulation and better control after optimization. Ongoing exploration of cyanobacterial transcription factors and stress-responsive regulatory networks is expected to drive the development of more precise, tunable, and efficient inducible promoter systems for advanced gene regulation.

- Constitutive promoters

  Constitutive promoters drive continuous gene expression independent of external stimuli and are essential tools for stable transgene expression in cyanobacteria. Several endogenous promoters, such as P*cpcB*, P*psbA2*, and P*rnpB*, have been widely used for heterologous expression [145, 155, 168]. However, many native promoters are influenced by the circadian clock, resulting in variable expression across a 24-hour period [128, 169, 170]. This necessitates a focus on optimizing native promoters and developing synthetic variants to overcome such limitations. Truncation of native promoters has proven effective; for instance, P*cpc560*, derived from P*cpcB* of *Synechocystis* PCC 6803, achieved protein levels up to 15% of total soluble proteins [135, 137, 159], while a truncated P*psbA2S* showed a fourfold increase in gene expression [71]. Similarly, a tandem hybrid promoter combining a truncated native PR from *Synechococcus* 7942 and the PS promoter from *E. coli* resulted in high expression efficiency [134].

  High-throughput and bioinformatics-based approaches, including error-prone PCR and RNA-seq analysis, have enabled the generation of promoter libraries with a broad range of strengths [96, 148, 152]. Moreover, computational tools such as Virtual Footprint and bTSSfinder are useful in identifying new strong endogenous promoters for cyanobacterial systems [171].

3.4 Ribosome binding sites

RBSs are short nucleotide sequences located in the 5′ untranslated region (UTR) of mRNAs that play a crucial role in initiating translation by recruiting ribosomes to the start codon. In polycistronic operons, each open reading frame (ORF) contains its own RBS to ensure translation of all encoded proteins [87]. The canonical Shine–Dalgarno (SD) sequence (5′-GGAGG-3′) pairs with the complementary anti-SD region of the 16S rRNA, positioning the start codon correctly within the ribosome [3, 87]. In cyanobacteria such as *Synechocystis* PCC 6803 and *Synechococcus elongatus* PCC 7942, both native and synthetic RBSs have been characterized and optimized for gene expression [87, 94]. For instance, the native 5′ UTR of P*cpc560* drives strong translation in *Synechocystis* [70], while synthetic sequences such as RBS* and BioBrick BBa_0034 provide high protein yields for fluorescent reporters [71]. Studies in *Synechocystis* and *S. elongatus* have revealed a broad range of RBS strengths, allowing modulation of gene expression at the translational level [6, 17, 71, 96, 139, 172, 173]. RBSs have also been used to coordinate multigene expression in synthetic operons or metabolic pathways. For example, in *S. elongatus* PCC 7942, differential RBS strengths were employed to balance expression of alsS, alsD, and adh, improving 2,3-butanediol production [174]. Similarly, optimization of RBS sequences enhanced ethylene (EFE) and limonene production in cyanobacteria [159, 175].

Computational design of RBSs has become an important strategy to fine-tune translation. Tools such as the RBS Calculator, UTR Designer, and RBS Designer use thermodynamic modeling to predict or design RBSs for desired translation rates [152, 176]. These tools show high predictive accuracy in *E. coli* ($R^2 > 0.8$) but perform less reliably in cyanobacteria, likely due to species-specific translational contexts [152, 172]. Despite the limitations, in silico RBS design combined with experimental validation has enabled substantial improvements in heterologous protein expression across cyanobacterial species such as *Anabaena* sp. PCC 7120, *Synechocystis* PCC 6803, and *S. elongatus* PCC 7942 [69, 71, 172]. Future research aims to develop cyanobacteria-specific RBS prediction tools to enhance accuracy and facilitate easier synthetic biology applications [139].

3.5 Riboswitches

Riboswitches are cis-acting regulatory RNA elements located mainly in the 5′ UTRs of mRNAs that modulate gene expression by changing their structure upon binding to specific ligands such as metabolites, ions, or signaling molecules. Each riboswitch consists of two parts: an aptamer domain, which binds the ligand, and an expression platform, which mediates gene regulation by influencing transcription, translation, or mRNA stability [177]. To date, at least 56 classes of natural riboswitches have been identified, responding to compounds such as thiamine pyrophosphate, cobalamin, glycine, lysine, Mg^{2+}, Mn^{2+}, c–di-GMP, and ppGpp [178].

In cyanobacteria, functionally characterized native riboswitches include the glutamine-I riboswitch, which regulates glutamine synthetase activity in *Synechocystis* PCC 6803 [151], and cobalamin-dependent riboswitches in *Nostoc* sp. PCC 7120 [179] and *Synechococcus* PCC 7002 [180]. These native elements participate in nitrogen assimilation and vitamin B_{12}-related metabolism, but their limited ligand affinity and potential interference with essential metabolic pathways restrict their practical application [177, 181, 182]. To overcome these challenges, synthetic riboswitches have been developed to provide precise, tunable, and protein-factor-independent gene regulation [182]. Among them, the theophylline-dependent riboswitch, a modified RNA element that exposes or hides the RBS upon binding theophylline, has become the most widely used system in cyanobacteria [183–185]. Initially implemented in *Synechococcus* PCC 7942, it achieved up to 190-fold induction with negligible basal expression [183] and was later successfully applied to *Synechocystis* PCC 6803, *Anabaena* PCC 7120, *Leptolyngbya* BL0902, and *Synechocystis* WHSyn [118, 154, 186]. A cobalamin-dependent riboswitch from *Synechococcus* PCC 73109 has also been adapted for *Synechococcus* 7002, offering low leakiness and about six-fold induction [180]. However, its applicability is limited to cobalamin-auxotrophic species, since others can synthesize the cofactor internally [187].

Overall, riboswitches provide compact, tunable, and ligand-responsive RNA tools for gene regulation in cyanobacteria. Their independence from protein inducers, modularity, and fast response make them promising for metabolic engineering [188, 189]. Future work should aim at combining in silico structural design with experimental tuning to expand the number of functional riboswitches and enhance their versatility across diverse cyanobacterial species [190].

3.6 Marker and marker-less selection system

In cyanobacterial genetic engineering, marker-based selection typically involves inserting a gene of interest alongside an antibiotic resistance gene to identify successful transformants, and recombinants are confirmed using PCR, sequencing, or Southern blotting [10, 17]. However, for multiple or stable modifications, marker-less selection systems are preferred as they avoid the accumulation of antibiotic markers and allow repeated genome editing [87]. These systems often use a two-step process combining a positive selection marker (antibiotic resistance) and a negative selection marker (lethal gene) such as *sac*B, which produces toxic compounds in the presence of sucrose. Only cells that lose the *sac*B gene survive on sucrose, enabling the recovery of unmarked mutants [191]. This counter-selection strategy has been successfully applied in several cyanobacterial strains, including *Synechocystis* [17, 84, 192–195].

Early marker less systems in cyanobacteria relied on mutation-based selection, such as the rps12 system in *Synechococcus elongatus* 7942, which used a streptomycin-sensitive phenotype for double selection [196, 197]. Later, new counter-selection methods were developed, including those based on organic acid toxicity (acsA/acrylate system) in *Synechococcus* PCC 7002, nickel-inducible toxin expression (mazF system) in *Synechocystis* PCC 6803 [73], and nptI–sacB cassettes for single-vector deletions [198]. Some systems were found to be strain-specific, as *sac*B selection fails in marine cyanobacteria like *Synechococcus* PCC 7002 due to salt sensitivity [199, 200]. Broader methods, such as the aas knockout–based selection using free fatty acid tolerance, work across both *Synechocystis* PCC 6803 and *Synechococcus* PCC 7002 [201]. The most recent advancement in marker less genome editing uses CRISPR-based technologies, which create targeted DNA breaks without requiring counter-selection genes, thereby simplifying and accelerating the process [75, 76].

3.7 CRISPR-derived tools

A recent breakthrough in cyanobacterial genome engineering is the application of CRISPR-derived technologies. These approaches are generally classified into two main types: Genome editing and CRISPRi. Genome editing involves creating targeted double-strand breaks using a CRISPR-associated (Cas) endonuclease guided by a single-guide RNA, followed by homologous recombination to introduce new DNA sequences [202], while CRISPRi employs a catalytically inactive Cas protein, such as dead Cas9 (dCas9), to repress gene expression without cutting the DNA. Furthermore, cyanobacteria themselves are an abundant natural reservoir of diverse and novel CRISPR-Cas systems [203–205].

CRISPR-Cas–based genome editing enables a range of genetic modifications in cyanobacteria, including gene knock-ins, deletions, and site-specific point mutations. A recent advancement involves RNA-guided transposition using a CRISPR-associated transposase, successfully demonstrated in *Nostoc sp. PCC 7120* with Golden Gate vectors [206]. The introduction of CRISPR/Cas9 technology significantly enhanced the efficiency of homologous recombination in cyanobacteria, as first shown in *Synechococcus elongatus* strains PCC 7942 and UTEX 2973 [207, 208]. However, excessive Cas9 expression was found to be toxic to cells, a limitation that can be mitigated by employing Cas12a (formerly Cpf1) [74]. Cas12a differs from Cas9 in several important ways: it recognizes an AT-rich PAM sequence (5′-TTN-3′), which better suits the AT-rich nature of many cyanobacterial genomes; it can process precursor crRNAs into mature crRNAs autonomously, without the need for host RNase; and it introduces staggered double-strand breaks, thereby improving recombination efficiency [209]. The effectiveness of Cas12a-mediated editing in cyanobacteria was demonstrated in *S. elongatus* UTEX 2973, where genetic variations between this fast-growing strain and the slower-growing *S. elongatus* PCC 7942 were systematically analyzed. In that study, the pUH24 plasmid was first eliminated from UTEX 2973, and each of the 39 chromosomal SNPs was individually converted to its corresponding allele in PCC 7942 using the Cas12a system, highlighting its precision and robustness [210]. The Cas12a-based CRISPR system has been adapted for cyanobacterial genome editing through the pSEVA plasmid series, which utilizes the broad-host-range RSF1010 origin [211]. This system has been successfully applied across various cyanobacterial species and strains, including *Synechocystis sp.*, *Synechococcus elongatus* PCC 7942, UTEX 2973, and 11081, *Anabaena (Nostoc) sp. PCC 7120*, and *Chroococcidiopsis sp.* B13 [74, 76, 211, 212]. To mitigate Cas9-associated toxicity, another strategy involves regulating its expression using a theophylline-responsive riboswitch, which allows precise and conditional control of Cas9 activity. This approach has been proven effective for CRISPR/Cas9-mediated genome manipulations in *Synechocystis*, including the introduction of insertions and deletions of varying lengths [213]. This is extensively reviewed in the literature [75, 214].

CRISPRi uses catalytically inactive enzymes, such as dCas9 or dCas12a, which bind to DNA without cutting it, allowing for the repression of essential genes. The first demonstration of CRISPRi in cyanobacteria was in *Synechocystis sp. PCC 6803*, where a gene encoding green fluorescent protein (GFP) was successfully silenced [215], followed by the repression of essential genes like *rne*, encoding endoribonuclease E [75]. Since then, CRISPRi has been applied for metabolic engineering in several species, including *Synechocystis*, *S. elongatus* PCC 7942, *Synechococcus sp.* PCC 7002, and *Nostoc sp.* PCC 7120 [9, 72, 153, 216].

Since CRISPRi causes partial repression rather than complete knockout, it enables genome-wide studies using single-guide RNA (sgRNA) libraries. For instance, an inducible CRISPRi library in *Synechocystis* targeting over 3,500 protein-coding genes and 1,800 noncoding RNA (ncRNA) loci was developed [217]. This approach allowed high-throughput gene fitness analysis under various growth conditions, helping identify key genes and enzymes with dual metabolic functions. The system was further used to improve L-lactate production, where repression of genes such as *gltA* (citrate synthase) and *pcnB* (CCA-tRNA nucleotidyl transferase) enhanced yield [217]. Additional targets like *sdhB* and *ssr2016*, involved in redox balance and cyclic electron flow, also correlated with improved productivity under controlled light conditions.

Interestingly, CRISPRi screens also revealed that several ncRNAs influence fitness and metabolic regulation, highlighting their potential roles in fine-tuning cyanobacterial metabolism [218–221]. One such regulatory RNA, ApcZ, derived from the 3′ UTR of the *apcC* gene, links light harvesting with energy dissipation by modulating orange carotenoid protein expression. These findings emphasize CRISPRi's potential not only for targeted gene repression but also for uncovering complex regulatory networks in cyanobacteria.

4. Applications of genetic engineering in cyanobacteria

Genetic engineering of cyanobacteria has opened up vast possibilities for sustainable biotechnological applications, driven by their ability to convert solar energy and atmospheric CO_2 into valuable bioproducts [159]. Engineered cyanobacteria have been explored as cell factories for the biosynthesis of renewable fuels, chemicals, and high-value proteins [159]. Through the integration of synthetic biology tools, metabolic pathways have been redesigned to enhance the production of compounds such as ethanol, hydrogen, isoprene, lactate, and 2,3-butanediol, positioning cyanobacteria as promising green biofactories [4–7, 159]. Recent advances have also extended to protein production, where cyanobacteria are engineered to express recombinant enzymes, pharmaceuticals, and nutraceuticals directly from atmospheric nitrogen and CO_2 [222]. Such systems provide a renewable alternative to conventional protein sources like soy or animal feed, reducing land and fertilizer dependency while minimizing greenhouse gas emissions [222]. Cyanobacterial strains have been optimized for industrial protein bioreactors, offering high yields under controlled photobioreactor conditions [104, 223]. Moreover, cyanobacteria are being exploited for the biosynthesis of bioactive metabolites, including pigments, carotenoids, polysaccharides, and antioxidants, with potential applications in the pharmaceutical, cosmetic, and nutraceutical industries [12, 224]. Genetic and metabolic engineering approaches have enhanced these pathways for improved yield and specificity [10, 11, 225]. Synthetic gene designs combining engineered promoters, RBSs, and terminators now enable fine control over protein expression and carbon flux, enhancing the production of amino acids and storage proteins [9, 159].

Together, these developments position cyanobacteria as a powerful platform for sustainable biotechnology. By combining metabolic engineering, CRISPR-based genome editing, and synthetic biology design principles, modern cyanobacterial systems can produce renewable biofuels, high-value chemicals, and nutritious proteins efficiently, while supporting carbon-neutral industrial processes [159, 222].

5. Future prospects

The progressive development of genetic tools in cyanobacteria reflects one of the most remarkable transitions in microbial biotechnology, from basic physiological studies to precise genome engineering and pathway optimization. Each decade brought a critical advance: the establishment of transformation systems in the 1980s, the construction of shuttle vectors and selectable markers in the 1990s, genome sequencing in the late 1990s, and the emergence of synthetic biology and CRISPR/Cas technologies in the twenty-first century. Together, these milestones have transformed cyanobacteria into robust model organisms capable of supporting complex genetic designs. Modern toolkits now allow controlled expression of heterologous and synthetic pathways, marker-less genome modifications, and orthogonal regulatory systems compatible with photosynthetic metabolism. Meanwhile, the future of cyanobacterial genetic engineering is set to be revolutionized by the integration of artificial intelligence (AI), computational biology, and multiomics technologies, which together will redefine how cyanobacterial strains are designed and optimized. AI and machine learning (ML) models are increasingly being used to analyze large biological datasets and predict how DNA sequence variations influence gene expression and metabolic output [226, 227]. These algorithms can help identify strong promoters or RBSs, predict transcriptional efficiency, and design new synthetic regulatory elements tailored for cyanobacteria [228, 229]. Integrated within the Design–Build–Test–Learn (DBTL) cycle, AI-based models enable predictive design and refinement of genetic constructs, significantly reducing the time and cost associated with experimental optimization [227].

Multiomics-guided strain optimization, which integrates genomics, transcriptomics, proteomics, and metabolomics, has also become an important part of modern genetic engineering. These datasets feed into genome-scale metabolic models (GSMs) that simulate cellular metabolism and identify genetic targets for improving productivity [229]. For instance, AI-driven omics integration in *Synechocystis sp.* PCC 6803 enabled the reconstruction of regulatory networks that enhanced photosynthetic efficiency and redirected carbon flux towards desired metabolites [230]. Similarly, systems biology-based modeling in *Synechococcus elongatus* revealed metabolic responses to environmental stress, guiding the engineering of strains capable of producing higher yields of glycogen and fatty acids [231]. Thus, AI and multiomics integration can bridge the gap between computational predictions and experimental outcomes, making strain improvement more rational and data-driven.

Additionally, cell-free systems and minimal cyanobacterial chassis are emerging as powerful tools to overcome the inherent complexity of whole-cell systems. Cell-free systems are cellular extracts containing transcription and translation machinery, which allow precise and rapid testing of genetic constructs without the constraints of cellular regulation [232]. Minimal cyanobacterial chassis, such as *Synechococcus elongatus* UTEX 2973, are being developed as simplified yet highly efficient production hosts due to their fast growth and robust photosynthetic capacity [233]. These simplified systems make results more consistent, reduce unwanted cell activity, and make it easier to predict gene behavior.

Despite these innovations, the use of synthetic biology tools and strategies for designing and optimizing light-induced cyanobacterial cell factories is still at an early stage. Major challenges for future research include enhancing the efficiency of genetic tools, developing robust and versatile cyanobacterial chassis, simplifying and fine-tuning metabolic regulation, and advancing as well as integrating metabolic modeling approaches [223]. Additionally, scaling up genetic engineering for industrial biotechnology is not easy, as engineered strains often exhibit instability, resource competition, and reduced productivity under variable outdoor conditions. Large-scale cultivation also faces limitations such as uneven light penetration, restricted CO_2 diffusion, and biofilm formation [230, 234]. Overcoming these barriers will require more research focused on integrating AI-driven predictive design, multiomics data, and cell-free prototyping with advanced bioprocess engineering and evolution-based strain improvement. These approaches will help in developing robust, scalable, and sustainable cyanobacterial biofactories capable of meeting the growing global demand for renewable fuels, feedstocks, and high-value biochemicals.

Acknowledgments

Miss Kritika Tripathi is thankful to the University Grants Commission (UGC) for JRF and SRF.

Conflict of Interest

The authors declare no conflict of interest.

Author details

Kritika Tripathi and Yogesh Mishra*
Department of Botany, Institute of Science, Banaras Hindu University (BHU), Varanasi, India

*Address all correspondence to: ymishra@bhu.ac.in

References

[1] Lanigan TM, Kopera HC, Saunders TL. Principles of genetic engineering. Genes. 2020;**11**(3):291

[2] Diamond J. Evolution, consequences and future of plant and animal domestication. Nature. 2002;**418**(6898):700–707

[3] Gordon GC, Pfleger BF. Regulatory tools for controlling gene expression in cyanobacteria. Synthetic Biology of Cyanobacteria. Vol. **1080**. Singapore: Springer; 2018. p. 281–315

[4] Sakai M, Ogawa T, Matsuoka M, Fukuda H. Photosynthetic conversion of carbon dioxide to ethylene by the recombinant cyanobacterium, Synechococcus sp. PCC 7942, which harbors a gene for the ethylene-forming enzyme of Pseudomonas syringae. Journal of Fermentation and Bioengineering. 1997;**84**(5):434–443

[5] Niederholtmeyer H, Wolfstädter BT, Savage DF, Silver PA, Way JC. Engineering cyanobacteria to synthesize and export hydrophilic products. Applied and Environmental Microbiology. 2010;**76**(11):3462–3466

[6] Oliver JW, Machado IM, Yoneda H, Atsumi S. Cyanobacterial conversion of carbon dioxide to 2, 3-butanediol. Proceedings of the National Academy of Sciences. 2013;**110**(4):1249–1254

[7] Wang B, Wang J, Zhang W, Meldrum DR. Application of synthetic biology in cyanobacteria and algae. Frontiers in Microbiology. 2012;**3**:344

[8] Al-Amin A, Parvin F, Chakraborty J, Kim YI. Cyanobacteria mediated heavy metal removal: A review on mechanism, biosynthesis, and removal capability. Environmental Technology Reviews. 2021;**10**(1):44–57

[9] Xue Y, He Q. Cyanobacteria as cell factories to produce plant secondary metabolites. Front Bioeng Biotechnol. 2015;**3**:57. DOI: 10.3389/fbioe.2015.00057

[10] Wang F, Gao Y, Yang G. Recent advances in synthetic biology of cyanobacteria for improved chemicals production. Bioengineered. 2020;**11**(1):1208–1220

[11] Wang X, Chen L, Liu J, Sun T, Zhang W. Light-driven biosynthesis of myo-Inositol directly from CO2 in Synechocystis sp. PCC 6803. Frontiers in Microbiology. 2020;**11**: 566117

[12] Pagels F, Vasconcelos V, Guedes AC. Carotenoids from cyanobacteria: Biotechnological potential and optimization strategies. Biomolecules. 2021;**11**(5):735

[13] Watson JD, Crick FH. Molecular structure of nucleic acids: A structure for deoxyribose nucleic acid. Nature. 1953;**171**(4356):737–738

[14] Shen S, Loh TJ, Shen H, Zheng X, Shen H. CRISPR as a strong gene editing tool. BMB Reports. 2017;**50**(1):20

[15] Ramey CJ, Barón-Sola A, Aucoin HR, Boyle NR. Genome engineering in cyanobacteria: Where we are and where we need to go. ACS Synthetic Biology. 2015;**4**(11):1186–1196

[16] Stanier RY, Cohen-Bazire G. Phototrophic prokaryotes: The cyanobacteria. Annual Review of Microbiology. 1977;**31**(1):225–274

[17] Heidorn T, Camsund D, Huang HH, Lindberg P, Oliveira P, Stensjö K, Lindblad P. Synthetic biology in cyanobacteria: Engineering and analyzing novel functions. In: Voigt C, editor. Methods in Enzymology. Vol. **497**. Academic Press; 2011. p. 539–579

[18] Griese M, Lange C, Soppa J. Ploidy in cyanobacteria. FEMS Microbiology Letters. 2011;**323**(2):124–131

[19] Zerulla K, Ludt K, Soppa J. The ploidy level of Synechocystis sp. PCC 6803 is highly variable and is influenced by growth phase and by chemical and physical external parameters. Microbiology. 2016;**162**(5):730–739

[20] Moser DP, Zarka D, Kallas T. Characterization of a restriction barrier and electrotransformation of the cyanobacterium Nostoc PCC 7121. Archives of Microbiology. 1993;**160**(3):229–237

[21] Jeamton W, Dulsawat S, Tanticharoen M, Vonshak A, Cheevadhanarak S. Overcoming intrinsic restriction enzyme barriers enhances transformation efficiency in Arthrospira platensis C1. Plant and Cell Physiology. 2017;**58**(4):822–830

[22] Wang B, Yu J, Zhang W, Meldrum DR. Premethylation of foreign DNA improves integrative transformation efficiency in Synechocystis sp. strain PCC 6803. Applied and Environmental Microbiology. 2015;**81**(24):8500–8506

[23] Murphy RC, Bryant DA, Porter RD, De Marsac NT. Molecular cloning and characterization of the recA gene from the cyanobacterium Synechococcus sp. strain PCC 7002. Journal of Bacteriology. 1987;**169**(6):2739–2747

[24] Dong G, Golden SS. How a cyanobacterium tells time. Current Opinion in Microbiology. 2008;**11**(6):541–546

[25] Gale GA, Wang B, McCormick AJ. Evaluation and comparison of the efficiency of transcription terminators in different cyanobacterial species. Frontiers in Microbiology. 2021;**11**:624011

[26] Huang HH, Lindblad P. Wide-dynamic-range promoters engineered for cyanobacteria. Journal of Biological Engineering. 2013;7(1):10

[27] Flores E, Muro-Pastor AM, Meeks JC. Gene Transfer to cyanobacteria in the Laboratory and in Nature. In: Herrero A, Flores E, editors. The Cyanobacteria: Molecular Biology, Genomics and Evolution. Norfolk: Caiser Academic; 2008. p. 45–57

[28] Angermayr SA, Hellingwerf KJ, Lindblad P, de Mattos MJ. Energy biotechnology with cyanobacteria. Current Opinion in Biotechnology. 2009;**20**(3):257–263

[29] Gellert M, Little JW, Oshinsky CK, Zimmerman SB. Joining of DNA strands by DNA ligase of E. coli. In: Cold Spring Harbor Symposia on Quantitative Biology. Vol. **33**. Cold Spring Harbor, NY, USA: Cold Spring Harbor Laboratory Press; 1968. p. 21–26

[30] Arber W, Linn S. DNA modification and restriction. Annual Review of Biochemistry. 1969;**38**(1):467–500

[31] Smith HO, Welcox KW. A restriction enzyme from Hemophilus influenzae: I. Purification and general properties. Journal of Molecular Biology. 1970;**51**(2):379–391

[32] Cohen SN, Chang AC, Boyer HW, Helling RB. Construction of biologically functional bacterial plasmids in vitro. Proceedings of the National Academy of Sciences. 1973;**70**(11):3240–3244

[33] Mullis K, Faloona F, Scharf S, Saiki R, Horn G, Erlich H. Specific enzymatic amplification of DNA in vitro: The polymerase chain reaction. In: Cold Spring Harbor Symposia on Quantitative Biology. Vol. **51**. Cold Spring Harbor, NY, USA: Cold Spring Harbor Laboratory Press; 1986. p. 263–273

[34] Shestakov SV, Khyen NT. Evidence for genetic transformation in blue-green alga Anacystis nidulans. Molecular and General Genetics MGG. 1970;**107**(4):372–375

[35] Asato Y, Ginoza HS. Separation of Small Circular DNA Molecules from the Blue–green Alga Anacystis nidulans. Nature New Biology. 1973;**244** (135):132–133

[36] Hall G, Flick MB, Jensen RA. Approach to recognition of regulatory mutants of cyanobacteria. Journal of Bacteriology. 1980;**143** (2):981–988

[37] Wolk CP, Vonshak A, Kehoe P, Elhai J. Construction of shuttle vectors capable of conjugative transfer from Escherichia coli to nitrogen-fixing filamentous cyanobacteria. Proceedings of the National Academy of Sciences. 1984;**81**(5):1561–1565

[38] Elhai J, Wolk CP. Conjugal transfer of DNA to cyanobacteria. In: Packer L, Glazer AN, editors. Methods in Enzymology. Vol. **167**. Academic Press; 1988. p. 747–754

[39] Thiel TE, Poo HA. Transformation of a filamentous cyanobacterium by electroporation. Journal of Bacteriology. 1989;**171**(10):5743–5746

[40] Porter RD. Transformation in cyanobacteria. Critical Reviews in Microbiology. 1986;**13**(2):111–132

[41] De Marsac NT, Borrias WE, Kuhlemeier CJ, Castets AM, Van Arkel GA, Van den Hondel CA. A new approach for molecular cloning in cyanobacteria: Cloning of an Anacystis nidulans met gene using a Tn 907-induced mutant. Gene. 1982;**20** (1):111–119

[42] Borthakur DU, Haselkorn RO. Tn5 mutagenesis of Anabaena sp. strain PCC 7120: Isolation of a new mutant unable to grow without combined nitrogen. Journal of Bacteriology. 1989;**171** (10):5759–5761

[43] Wolk CP, Cai Y, Panoff JM. Use of a transposon with luciferase as a reporter to identify environmentally responsive genes in a cyanobacterium. Proceedings of the National Academy of Sciences. 1991;**88**(12):5355–5359

[44] Golden SS, Brusslan J, Haselkorn R. Genetic engineering of the cyanobacterial chromosome. In: Wu R, Grossman L, editors. Methods in Enzymology. Vol. **153**. Academic Press; 1987. p. 215–231

[45] Williams JG. Construction of specific mutations in photosystem II photosynthetic reaction center by genetic engineering methods in Synechocystis 6803. In: Packer L, Glazer AN, editors. Methods in Enzymology. Vol. **167**. Academic Press; 1988. p. 766–778

[46] Wolk CP, Cai Y, Cardemil L, Flores E, Hohn B, Murry M, Schmetterer G, Schrautemeier B, Wilson R. Isolation and

complementation of mutants of Anabaena sp. strain PCC 7120 unable to grow aerobically on dinitrogen. Journal of Bacteriology. 1988;**170**(3):1239–1244

[47] Haselkom R. Genetic systems in cyanobacteria. In: Jeffrey HM, editor. Methods in Enzymology. Vol. **204**. Academic Press; 1991. p. 418–430

[48] Ferino F, Chauvat F. A promoter-probe vector-host system for the cyanobacterium, Synechocystis PCC 6803. Gene. 1989;**84**(2):257–266

[49] Vachhani AK, Iyer RK, Tuli R. A mobilizable shuttle vector for the cyanobacterium Plectonema boryanum. Microbiology. 1993;**139**(3):569–573

[50] Thiel T. Genetic analysis of cyanobacteria In: Bryant DA, editor. The Molecular Biology of Cyanobacteria. Vol. **1**. Dordrecht: Springer; 1994. p. 581–611

[51] Houmard J, de Marsac NT. Cyanobacterial genetic tools: Current status. In: Packer L, Glazer AN, editors. Methods in Enzymology. Vol. **167**. Academic Press; 1988. p. 808–847

[52] Friedberg D, Seijffers J. A new hybrid plasmid capable of transforming Escherichia coli and Anacystis nidulans. Gene. 1983;**22**(2–3):267–275

[53] Thiel TE. Characterization of genes for an alternative nitrogenase in the cyanobacterium Anabaena variabilis. Journal of Bacteriology. 1993;**175**(19):6276–6286

[54] Bagdasarian M, Timmis KN. Host: vector systems for gene cloning in Pseudomonas. Current Topics in Microbiology and Immunology. 1982;**96**:47–67

[55] Kreps S, Ferino F, Mosrin C, Gerits J, Mergeay M, Thuriaux P. Conjugative transfer and autonomous replication of a promiscuous IncQ plasmid in the cyanobacterium Synechocystis PCC 6803. Molecular and General Genetics. 1990;**221**(1):129–133

[56] Sode K, Tatara M, Ogawa S, Matsunaga T. Maintenance of broad host range vector pKT230 in marine unicellular cyanobacteria. FEMS Microbiology Letters. 1992;**99**(1):73–78

[57] Elhai J, Wolk CP. A versatile class of positive-selection vectors based on the nonviability of palindrome-containing plasmids that allows cloning into long polylinkers. Gene. 1988;**68**(1):119–138

[58] Borthakur D, Basche M, Buikema WJ, Borthakur PB, Haselkorn R. Expression, nucleotide sequence and mutational analysis of two open reading frames in the nif gene region of Anabaena sp. strain PCC 7120. Molecular and General Genetics. 1990;**221**(2):227–234

[59] Elhai J, Wolk CP. Developmental regulation and spatial pattern of expression of the structural genes for nitrogenase in the cyanobacterium Anabaena. The EMBO Journal. 1990;**9**(10):3379–3388

[60] Scanlan DJ, Bloye SA, Mann NH, Hodgson DA, Carr NG. Construction of lacZ promoter probe vectors for use in Synechococcus: Application to the identification of CO2-regulated promoters. Gene. 1990;**90**(1):43–49

[61] Elhai J. Strong and regulated promoters in the cyanobacterium Anabaena PCC 7120. FEMS Microbiology Letters. 1993;**114**(2):179–184

[62] Schneider GJ, Lang JD, Haselkorn R. Promoter recognition by the RNA polymerase from vegetative cells of the cyanobacterium Anabaena 7120. Gene. 1991;**105**(1):51–60

[63] Friedberg D, Seijffers J. Controlled gene expression utilising lambda phage regulatory signals in a cyanobacterium host. Molecular and General Genetics. 1986;**203**(3):505–510

[64] Wolk CP, Elhai J, Kuritz T, Holland D. Amplified expression of a transcriptional pattern formed during development of Anabaena. Molecular Microbiology. 1993;**7**(3):441–445

[65] Kuritz T, Wolk CP. Expression of foreign genes in filamentous cyanobacteria. Journal of Phycology. 1993;**29**9308131

[66] Kaneko T, Sato S, Kotani H, Tanaka A, Asamizu E, Nakamura Y, Miyajima N, Hirosawa M, Sugiura, Sasamoto S, Kimura T. Sequence analysis of the genome of the unicellular cyanobacterium Synechocystis sp. strain PCC6803. II. Sequence determination of the entire genome and assignment of potential protein-coding regions. DNA Research. 1996;**3**(3):109–136

[67] Nakamura Y, Kaneko T, Hirosawa M, Miyajima N, Tabata S. CyanoBase, a www database containing the complete nucleotide sequence of the genome of Synechocystis sp. strain PCC6803. Nucleic Acids Research. 1998;**26**(1):63–67

[68] Knight T. Idempotent vector design for standard assembly of biobricks. 2003.

[69] Taton A, Unglaub F, Wright NE, Zeng WY, Paz-Yepes J, Brahamsha B, Palenik B, Peterson TC, Haerizadeh F, SS G, JW G. Broad-host-range vector system for synthetic biology and biotechnology in cyanobacteria. Nucleic Acids Research. 2014;**42**(17): e136

[70] Vasudevan R, Gale GA, Schiavon AA, Puzorjov A, Malin J, Gillespie MD, Vavitsas K, Zulkower V, Wang B, Howe CJ, Lea-Smith DJ. CyanoGate: A modular cloning suite for engineering cyanobacteria based on the plant MoClo syntax. Plant Physiology. 2019;**180**(1):39–55

[71] Englund E, Liang F, Lindberg P. Evaluation of promoters and ribosome binding sites for biotechnological applications in the unicellular cyanobacterium Synechocystis sp. PCC 6803. Scientific Reports. 2016;**6** (1):36640

[72] Huang CH, Shen CR, Li H, Sung LY, Wu MY, Hu YC. CRISPR interference (CRISPRi) for gene regulation and succinate production in cyanobacterium S. elongatus PCC 7942. Microbial Cell Factories. 2016;**15** (1):196

[73] Cheah YE, Albers SC, Peebles CA. A novel counter-selection method for markerless genetic modification in Synechocystis sp. PCC 6803. Biotechnology Progress. 2013;**29**(1): 23–30

[74] Ungerer J, Pakrasi HB. Cpf1 is a versatile tool for CRISPR genome editing across diverse species of cyanobacteria. Scientific Reports. 2016;**6**(1):39681

[75] Behler J, Vijay D, Hess WR, Akhtar MK. CRISPR-based technologies for metabolic engineering in cyanobacteria. Trends in Biotechnology. 2018;**36**(10):996–1010

[76] Niu TC, Lin GM, Xie LR, Wang ZQ, Xing WY, Zhang JY, Zhang CC. Expanding the potential of CRISPR-Cpf1-based genome editing technology in the cyanobacterium Anabaena PCC 7120. ACS Synthetic Biology. 2018;**8**(1):170–180

[77] Srivastava A, Kumar A, Biswas S, Srivastava V, Rajaram H, Mishra Y. Regulatory role of LexA in modulating photosynthetic redox poise and cadmium stress tolerance in the cyanobacterium, Anabaena sp. PCC7120. Environmental and Experimental Botany. 2022;**195**:104790

[78] Srivastava A, Kumar A, Biswas S, Kumar R, Srivastava V, Rajaram H, Mishra Y. Gamma (γ)-radiation stress response of the cyanobacterium Anabaena sp. PCC7120: Regulatory role of LexA and photophysiological changes. Plant Science. 2023;**326**:111529

[79] Kalwani P, Rath D, Ballal A. Loss of 2-Cys-Prx affects cellular ultrastructure, disturbs redox poise and impairs photo synthesis in cyanobacteria. Plant, Cell & Environment. 2022;**45**(10):2972–2986

[80] Segall-Shapiro TH, Sontag ED, Voigt CA. Engineered promoters enable constant gene expression at any copy number in bacteria. Nature Biotechnology. 2018;**36**(4):352–358

[81] Kunert A, Hagemann M, Erdmann N. Construction of promoter probe vectors for Synechocystis sp. PCC 6803 using the light-emitting reporter systems Gfp and LuxAB. Journal of Microbiological Methods. 2000;**41**(3):185–194

[82] Farasat I, Kushwaha M, Collens J, Easterbrook M, Guido M, Salis HM. Efficient search, mapping, and optimization of multi-protein genetic systems in diverse bacteria. Molecular Systems Biology. 2014;**10**(6):731

[83] Varman AM, Xiao Y, Pakrasi HB, Tang YJ. Metabolic engineering of Synechocystis sp. strain PCC 6803 for isobutanol production. Applied and Environmental Microbiology. 2013;**79**(3):908–914

[84] Andersson CR, Tsinoremas NF, Shelton J, Lebedeva NV, Yarrow J, Min H, Golden SS. Application of bioluminescence to the study of circadian rhythms in cyanobacteria. In: Abelson JN, Ziegler MM, Simon MI, Baldwin TO, editors. Methods in Enzymology. Vol. **305**. Academic Press; 2000. p. 527–542

[85] Peterson ES, McCue LA, Schrimpe-Rutledge AC, Jensen JL, Walker H, Kobold MA, Webb SR, Payne SH, Ansong C, Adkins JN, Cannon WR. VESPA: Software to facilitate genomic annotation of prokaryotic organisms through integration of proteomic and transcriptomic data. BMC Genomics. 2012;**13**(1):131

[86] Nagy C, Thiel K, Mulaku E, Mustila H, Tamagnini P, Aro EM, Pacheco CC, Kallio P. Comparison of alternative integration sites in the chromosome and the native plasmids of the cyanobacterium Synechocystis sp. PCC 6803 in respect to expression efficiency and copy number. Microbial Cell Factories. 2021;**20**(1):130

[87] Opel F, Axmann IM, Klähn S. The molecular toolset and techniques required to build cyanobacterial cell factories. In: Bühler K, Lindberg P, editors. Vol. **183**. Cyanobacteria in

Biotechnology: Applications and Quantitative Perspectives. Cham: Springer International Publishing; 2022. p. 65–103

[88] Schirmacher AM, Hanamghar SS, Zedler JA. Function and benefits of natural competence in cyanobacteria: From ecology to targeted manipulation. Life. 2020;**10**(11):249

[89] Shintani M, Sanchez ZK, Kimbara K. Genomics of microbial plasmids: Classification and identification based on replication and transfer systems and host taxonomy. Frontiers in Microbiology. 2015;**6**:242

[90] Norrander J, Kempe T, Messing J. Construction of improved M13 vectors using oligodeoxynucleotide-directed mutagenesis. Gene. 1983;**26**(1):101–106

[91] Guerry P, Van Embden J, Falkow S. Molecular nature of two nonconjugative plasmids carrying drug resistance genes. Journal of Bacteriology. 1974;**117**(2):619–630

[92] Bishé B, Taton A, Golden JW. Modification of RSF1010-based broad-host-range plasmids for improved conjugation and cyanobacterial bioprospecting. Iscience. 2019;**20**:216–228

[93] Klähn S, Opel F, Hess WR. Customized molecular tools to strengthen metabolic engineering of cyanobacteria. Green Carbon. 2024;**2**(2):149–163

[94] Santos-Merino M, Singh AK, Ducat DC. New applications of synthetic biology tools for cyanobacterial metabolic engineering. Frontiers in Bioengineering and Biotechnology. 2019;**7**:33

[95] Opel F, Siebert NA, Klatt S, Tüllinghoff A, Hantke JG, Toepel J, Bühler B, Nürnberg DJ, Klähn S. Generation of synthetic shuttle vectors enabling modular genetic engineering of cyanobacteria. ACS Synthetic Biology. 2022;**11**(5):1758–1771

[96] Liu D, Pakrasi HB. Exploring native genetic elements as plug-in tools for synthetic biology in the cyanobacterium Synechocystis sp. PCC 6803. Microbial Cell Factories. 2018;**17**(1):48

[97] Jin H, Wang Y, Idoine A, Bhaya D. Construction of a shuttle vector using an endogenous plasmid from the cyanobacterium Synechocystis sp. PCC6803. Frontiers in Microbiology. 2018;**9**:1662

[98] Xiao Y, Wang S, Rommelfanger S, Balassy A, Barba-Ostria C, Gu P, Galazka JM, Zhang F. Developing a Cas9-based tool to engineer native plasmids in Synechocystis sp. PCC 6803. Biotechnology and Bioengineering. 2018;**115**(9):2305–2314

[99] Sakamaki Y, Maeda K, Nimura-Matsune K, Chibazakura T, Watanabe S. Characterization of a cyanobacterial rep protein with broad-host range and its utilization for expression vectors. Frontiers in Microbiology. 2023;**14**:1111979

[100] Kaltenbrunner A, Reimann V, Hoffmann UA, Aoyagi T, Sakata M, Nimura-Matsune K, Watanabe S, Steglich C, Wilde A, Hess WR. Regulation of pSYSA defense plasmid copy number in Synechocystis through RNase E and a highly transcribed asRNA. Frontiers in Microbiology. 2023;**14**:1112307

[101] Ohdate K, Sakata M, Maeda K, Sakamaki Y, Nimura-Matsune K,

Ohbayashi R, Hess WR, Watanabe S. Discovery of novel replication proteins for large plasmids in cyanobacteria and their potential applications in genetic engineering. Frontiers in Microbiology. 2024;**15**:1311290

[102] Sherman LA, Van de Putte P. Construction of a hybrid plasmid capable of replication in the bacterium Escherichia coli and the cyanobacterium Anacystis nidulans. Journal of Bacteriology. 1982;**150**(1):410–413

[103] Golden SS, Sherman LA. A hybrid plasmid is a stable cloning vector for the cyanobacterium Anacystis nidulans R2. Journal of Bacteriology. 1983;**155**(3):966–972

[104] Chen Y, Taton A, Go M, London RE, Pieper LM, Golden SS, Golden JW. Self-replicating shuttle vectors based on pANS, a small endogenous plasmid of the unicellular cyanobacterium Synechococcus elongatus PCC 7942. Microbiology. 2016;**162**(12):2029–2041

[105] Buzby JS, Porter RD, Stevens Jr SE. Expression of the Escherichia coli lacZ gene on a plasmid vector in a cyanobacterium. Science. 1985;**230**(4727):805–807

[106] Wolk CP, Fan Q, Zhou R, Huang G, Lechno-Yossef S, Kuritz T, Wojciuch E. Paired cloning vectors for complementation of mutations in the cyanobacterium Anabaena sp. strain PCC 7120. Archives of Microbiology. 2007;**188**(6):551–563

[107] Buikema WJ, Haselkorn RO. Isolation and complementation of nitrogen fixation mutants of the cyanobacterium Anabaena sp. strain PCC 7120. Journal of Bacteriology. 1991;**173**(6):1879–1885

[108] Stucken K, Ilhan J, Roettger M, Dagan T, Martin WF. Transformation and conjugal transfer of foreign genes into the filamentous multicellular cyanobacteria (subsection V) Fischerella and Chlorogloeopsis. Current Microbiology. 2012;**65**(5):552–560

[109] Billi D, Friedmann EI, Helm RF, Potts M. Gene transfer to the desiccation-tolerant cyanobacterium Chroococcidiopsis. Journal of Bacteriology. 2001;**183**(7):2298–2305

[110] Ravindran CM, Suguna S, Shanmugasundaram S. Electroporation as a tool to transfer the plasmid pRL489 in Oscillatoria MKU 277. Journal of Microbiological Methods. 2006;**66**(1):174–176

[111] Summers ML, Wallis JG, Campbell EL, Meeks JC. Genetic evidence of a major role for glucose-6-phosphate dehydrogenase in nitrogen fixation and dark growth of the cyanobacterium Nostoc sp. strain ATCC 29133. Journal of Bacteriology. 1995;**177**(21):6184–6194

[112] Argueta C, Yuksek K, Summers M. Construction and use of GFP reporter vectors for analysis of cell-type-specific gene expression in Nostoc punctiforme. Journal of Microbiological Methods. 2004;**59**(2):181–188

[113] Walton DK, Gendel SM, Atherly AG. DNA sequence and shuttle vector construction of plasmid pGL3 from Plectonema boryanum PCC 6306. Nucleic Acids Research. 1993;**21**(3):746

[114] Cobley JG, Zerweck E, Reyes R, Mody A, Seludo-Unson JR, Jaeger H, Weerasuriya S, Navankasattusas S. Construction of shuttle plasmids which

can be efficiently mobilized from Escherichia coli into the chromatically adapting cyanobacterium, Fremyella diplosiphon. Plasmid. 1993;**30**(2):90–105

[115] Albers SC, Gallegos VA, Peebles CA. Engineering of genetic control tools in Synechocystis sp. PCC 6803 using rational design techniques. Journal of Biotechnology. 2015;**216**:36–46

[116] Immethun CM, Ng KM, DeLorenzo DM, Waldron-Feinstein B, Lee YC, Moon TS. Oxygen-responsive genetic circuits constructed in Synechocystis sp. PCC 6803. Biotechnology and Bioengineering. 2016;**113**(2):433–442

[117] Immethun CM, DeLorenzo DM, Focht CM, Gupta D, Johnson CB, Moon TS. Physical, chemical, and metabolic state sensors expand the synthetic biology toolbox for Synechocystis sp. PCC 6803. Biotechnology and Bioengineering. 2017;**114**(7):1561–1569

[118] Taton A, Ma AT, Ota M, Golden SS, Golden JW. NOT gate genetic circuits to control gene expression in cyanobacteria. ACS Synthetic Biology. 2017;**6**(12):2175–2182

[119] Cao YQ, Li Q, Xia PF, Wei LJ, Guo N, Li JW, Wang SG. AraBAD based toolkit for gene expression and metabolic robustness improvement in Synechococcus elongatus. Scientific Reports. 2017;7(1):18059

[120] Kelly CL, Taylor GM, Hitchcock A, Torres-Mendez A, Heap JT. A rhamnose-inducible system for precise and temporal control of gene expression in cyanobacteria. ACS Synthetic Biology. 2018;7(4):1056–1066

[121] Kunert A, Vinnemeier J, Erdmann N, Hagemann M. Repression by Fur is not the main mechanism controlling the iron-inducible isiAB operon in the cyanobacterium Synechocystis sp. PCC 6803. FEMS Microbiology Letters. 2003;**227**(2): 255–262

[122] Englund E, Andersen-Ranberg J, Miao R, Hamberger B, Lindberg P. Metabolic engineering of Synechocystis sp. PCC 6803 for production of the plant diterpenoid manoyl oxide. ACS Synthetic Biology. 2015;**4**(12): 1270–1278

[123] Peca L, Kos PB, Vass IJ. Characterization of the activity of heavy metal-responsive promoters in the cyanobacterium Synechocystis PCC 6803. Acta Biologica Hungarica. 2007;**58**(Suppl 1):11–22

[124] Peca L, Kós PB, Máté Z, Farsang A, Vass I. Construction of bioluminescent cyanobacterial reporter strains for detection of nickel, cobalt and zinc. FEMS Microbiology Letters. 2008;**289**(2):258–264

[125] Blasi B, Peca L, Vass I, Kós PB. Characterization of stress responses of heavy metal and metalloid inducible promoters in Synechocystis PCC6803. Journal of Microbiology and Biotechnology. 2012;**22**(2):166–169

[126] Tüllinghoff A, Uhl MB, Nintzel FE, Schmid A, Bühler B, Toepel J. Maximizing photosynthesis-driven Baeyer–Villiger oxidation efficiency in recombinant Synechocystis sp. PCC6803. Frontiers in Catalysis. 2022;**1**:780474

[127] Guerrero F, Carbonell V, Cossu M, Correddu D, Jones PR. Ethylene synthesis and regulated

expression of recombinant protein in Synechocystis sp. PCC 6803. PloS One. 2012;**7**(11)

[128] Camsund D, Lindblad P. Engineered transcriptional systems for cyanobacterial biotechnology. Frontiers in Bioengineering and Biotechnology. 2014;**2**:40

[129] Tan X, Yao L, Gao Q, Wang W, Qi F, Lu X. Photosynthesis driven conversion of carbon dioxide to fatty alcohols and hydrocarbons in cyanobacteria. Metabolic Engineering. 2011;**13**(2):169–176

[130] Oliveira P, Lindblad P. An AbrB-like protein regulates the expression of the bidirectional hydrogenase in Synechocystis sp. strain PCC 6803. Journal of Bacteriology. 2008;**190**(3):1011–1019

[131] Klähn S, Schaal C, Georg J, Baumgartner D, Knippen G, Hagemann M, Muro-Pastor AM, Hess WR. The sRNA NsiR4 is involved in nitrogen assimilation control in cyanobacteria by targeting glutamine synthetase inactivating factor IF7. Proceedings of the National Academy of Sciences. 2015;**112**(45):E6243–52

[132] Bolay P, Rozbeh R, Muro-Pastor MI, Timm S, Hagemann M, Florencio FJ, Forchhammer K, Klähn S. The novel PII-interacting protein PirA controls flux into the cyanobacterial ornithine-ammonia cycle. MBio. 2021;**12**(2):10–128

[133] Brandenburg F, Theodosiou E, Bertelmann C, Grund M, Klähn S, Schmid A, Krömer JO. Trans-4-hydroxy-L-proline production by the cyanobacterium Synechocystis sp. PCC 6803. Metabolic Engineering Communications. 2021;**12**:e00155

[134] Chungjatupornchai W, Fa-Aroonsawat S. The rrnA promoter as a tool for the improved expression of heterologous genes in cyanobacteria. Microbiological Research. 2014;**169** (5–6):361–368

[135] Ng AH, Berla BM, Pakrasi HB. Fine-tuning of photoautotrophic protein production by combining promoters and neutral sites in the cyanobacterium Synechocystis sp. strain PCC 6803. Applied and Environmental Microbiology. 2015;**81** (19):6857–6863

[136] Behle A, Saake P, Germann AT, Dienst D, Axmann IM. Comparative dose–response analysis of inducible promoters in cyanobacteria. ACS Synthetic Biology. 2020;**9** (4):843–855

[137] Zhou J, Zhang H, Meng H, Zhu Y, Bao G, Zhang Y, Li Y, Ma Y. Discovery of a super-strong promoter enables efficient production of heterologous proteins in cyanobacteria. Scientific Reports. 2014;**4**(1):4500

[138] Shono C, Ariyanti D, Abe K, Sakai Y, Sakamoto I, Tsukakoshi K, Sode K, Ikebukuro K. A green light-regulated T7 RNA polymerase gene expression system for cyanobacteria. Marine Biotechnology. 2021;**23**(1):31–38

[139] Wang B, Eckert C, Maness PC, Yu J. A genetic toolbox for modulating the expression of heterologous genes in the cyanobacterium Synechocystis sp. PCC 6803. ACS Synthetic Biology. 2018;**7**(1):276–286

[140] Dzelzkalns VA, Owens GC, Bogorad L. Chloroplast promoter driven expression of the chloramphenicol acetyl transferase gene in a

cyanobacterium. Nucleic Acids Research. 1984;**12**(23):8917–8925

[141] Jacobsen JH, Frigaard NU. Engineering of photosynthetic mannitol biosynthesis from CO2 in a cyanobacterium. Metabolic Engineering. 2014;**21**:60–70

[142] Nair U, Thomas C, Golden SS. Functional elements of the strong psbAI promoter of Synechococcus elongatus PCC 7942. Journal of Bacteriology. 2001;**183**(5):1740–1747

[143] Takahama K, Matsuoka M, Nagahama K, Ogawa T. Construction and analysis of a recombinant cyanobacterium expressing a chromosomally inserted gene for an ethylene-forming enzyme at the psbAI locus. Journal of Bioscience and Bioengineering. 2003;**95** (3):302–305

[144] Ducat DC, Sachdeva G, Silver PA. Rewiring hydrogenase-dependent redox circuits in cyanobacteria. Proceedings of the National Academy of Sciences. 2011;**108**(10):3941–3946

[145] Lindberg P, Park S, Melis A. Engineering a platform for photosynthetic isoprene production in cyanobacteria, using Synechocystis as the model organism. Metabolic Engineering. 2010;**12**(1):70–79

[146] Liu X, Sheng J, Curtiss III R. Fatty acid production in genetically modified cyanobacteria. Proceedings of the National Academy of Sciences. 2011;**108** (17):6899–6904

[147] Albers SC, Peebles CA. Evaluating light-induced promoters for the control of heterologous gene expression in Synechocystis sp. PCC 6803. Biotechnology Progress. 2017;**33**(1):45–53

[148] Ruffing AM, Jensen TJ, Strickland LM. Genetic tools for advancement of Synechococcus sp. PCC 7002 as a cyanobacterial chassis. Microbial Cell Factories. 2016;**15**(1):190

[149] Song K, Tan X, Liang Y, Lu X. The potential of Synechococcus elongatus UTEX 2973 for sugar feedstock production. Applied Microbiology and Biotechnology. 2016;**100**(18):7865–7875

[150] Yu J, Liberton M, Cliften PF, Head RD, Jacobs JM, Smith RD, Koppenaal DW, Brand JJ, Pakrasi HB. Synechococcus elongatus UTEX 2973, a fast-growing cyanobacterial chassis for biosynthesis using light and CO2. Scientific Reports. 2015; **5**(1):8132

[151] Klähn S, Bolay P, Wright PR, Atilho RM, Brewer KI, Hagemann M, Breaker RR, Hess WR. A glutamine riboswitch is a key element for the regulation of glutamine synthetase in cyanobacteria. Nucleic Acids Research. 2018;**46**(19):10082–10094

[152] Markley AL, Begemann MB, Clarke RE, Gordon GC, Pfleger BF. Synthetic biology toolbox for controlling gene expression in the cyanobacterium Synechococcus sp. strain PCC 7002. ACS Synthetic Biology. 2015;**4**(5):595–603

[153] Higo A, Isu A, Fukaya Y, Ehira S, Hisabori T. Application of CRISPR interference for metabolic engineering of the heterocyst-forming multicellular cyanobacterium Anabaena sp. PCC 7120. Plant and Cell Physiology. 2018;**59** (1):119–127

[154] Ma AT, Schmidt CM, Golden JW. Regulation of gene expression in diverse cyanobacterial species by using theophylline-responsive riboswitches. Applied and Environmental Microbiology. 2014;**80**(21):6704–6713

[155] Huang HH, Camsund D, Lindblad P, Heidorn T. Design and characterization of molecular tools for a synthetic biology approach towards developing cyanobacterial biotechnology. Nucleic Acids Research. 2010;**38**(8):2577–2593

[156] Mulo P, Sakurai I, Aro EM. Strategies for psbA gene expression in cyanobacteria, green algae and higher plants: From transcription to PSII repair. Biochimica Et Biophysica Acta (BBA) - Bioenergetics. 2012;**1817**(1):247–257

[157] Takahashi T, Nakai N, Muramatsu M, Hihara Y. Role of multiple HLR1 sequences in the regulation of the dual promoters of the psaAB genes in Synechocystis sp. PCC 6803. Journal of Bacteriology. 2010;**192**(15):4031–4036

[158] Wang HH, Kim H, Cong L, Jeong J, Bang D, Church GM. Genome-scale promoter engineering by coselection MAGE. Nature Methods. 2012;**9**(6):591–593

[159] Wang X, Liu W, Xin C, Zheng Y, Cheng Y, Sun S, Li R, Zhu XG, Dai SY, Rentzepis PM, Yuan JS. Enhanced limonene production in cyanobacteria reveals photosynthesis limitations. Proceedings of the National Academy of Sciences. 2016;**113**(50):14225–14230

[160] Jones PR. Genetic instability in cyanobacteria–an elephant in the room? Frontiers in Bioengineering and Biotechnology. 2014;**2**:12

[161] Hook-Barnard IG, Hinton DM. Transcription initiation by mix and match elements: Flexibility for polymerase binding to bacterial promoters. Gene Regulation and Systems Biology. 2007;**1**:117762500700100020

[162] Vijayan V, Jain IH, O'Shea EK. A high-resolution map of a cyanobacterial transcriptome. Genome Biology. 2011;**12**(5):R47

[163] Brahamsha B, Haselkorn R. Isolation and characterization of the gene encoding the principal sigma factor of the vegetative cell RNA polymerase from the cyanobacterium Anabaena sp. strain PCC 7120. Journal of Bacteriology. 1991;**173**(8):2442–2450

[164] Imamura S, Asayama M. Sigma factors for cyanobacterial transcription. Gene Regulation and Systems Biology. 2009;**3**:GRSB–S2090

[165] Berla BM, Saha R, Immethun CM, Maranas CD, Moon TS, Pakrasi HB. Synthetic biology of cyanobacteria: Unique challenges and opportunities. Frontiers in Microbiology. 2013;**4**:246

[166] Till P, Toepel J, Bühler B, RL M, Mach-Aigner AR. Regulatory systems for gene expression control in cyanobacteria. Applied Microbiology and Biotechnology. 2020;**104**(5):1977–1991

[167] Geerts D, Bovy A, de Vrieze G, Borrias M, Weisbeek P. Inducible expression of heterologous genes targeted to a chromosomal platform in the cyanobacterium Synechococcus sp PCC 7942. Microbiology. 1995;**141**(4):831–841

[168] Zhu T, Xie X, Li Z, Tan X, Lu X. Enhancing photosynthetic production of ethylene in genetically engineered Synechocystis sp. PCC 6803. Green Chemistry. 2015;**17** (1):421–434

[169] Liu Y, Tsinoremas NF, Johnson CH, Lebedeva NV, Golden SS, Ishiura M, Kondo T. Circadian orchestration of gene expression in cyanobacteria. Genes & Development. 1995;**9**(12):1469–1478

[170] Markson JS, Piechura JR, Puszynska AM, O'Shea EK. Circadian control of global gene expression by the cyanobacterial master regulator RpaA. Cell. 2013;**155**(6):1396–1408

[171] Shahmuradov IA, Mohamad Razali R, Bougouffa S, Radovanovic A, Bajic VB. bTSSfinder: A novel tool for the prediction of promoters in cyanobacteria and Escherichia coli. Bioinformatics. 2017;**33**(3):334–340

[172] Thiel K, Mulaku E, Dandapani H, etal. Translation efficiency of heterologous proteins is significantly affected by the genetic context of RBS sequences in engineered cyanobacterium Synechocystis sp. PCC 6803. Microbial Cell Factories. 2018;**17**:34

[173] Sebesta J, Peebles CA. Improving heterologous protein expression in Synechocystis sp. PCC 6803 for alpha-bisabolene production. Metabolic Engineering Communications. 2020;**10**: e00117

[174] Oliver JW, Machado IM, Yoneda H, Atsumi S. Combinatorial optimization of cyanobacterial 2, 3-butanediol production. Metabolic Engineering. 2014;**22**:76–82

[175] Xiong Q, Feng J, Li ST, Zhang GY, Qiao ZX, Chen Z, Wu Y, Lin Y, Li T, Ge F, Zhao JD. Integrated transcriptomic and proteomic analysis of the global response of Synechococcus to High light stress. Molecular & Cellular Proteomics. 2015;**14**(4): 1038–1053

[176] Reeve B, Hargest T, Gilbert C, Ellis T. Predicting translation initiation rates for designing synthetic biology. Frontiers in Bioengineering and Biotechnology. 2014;**2**:1

[177] Breaker RR. Prospects for riboswitch discovery and analysis. Molecular Cell. 2011;**43**(6):867–879

[178] Kavita K, Breaker RR. Discovering riboswitches: The past and the future. Trends in Biochemical Sciences. 2023;**48** (2):119–141

[179] Graf J, Fresenborg L, Seitz HM, Pernil R, Schleiff E. A cobalt concentration sensitive Btu-like system facilitates cobalamin uptake in Anabaena sp. PCC 7120. Microbial Cell. 2024;**11**:41

[180] Pérez AA, Liu Z, Rodionov DA, Li Z, Bryant DA. Complementation of cobalamin auxotrophy in Synechococcus sp. strain PCC 7002 and validation of a putative cobalamin riboswitch in vivo. Journal of Bacteriology. 2016;**198**(19):2743–2752

[181] Beisel CL, Smolke CD. Design principles for riboswitch function. PLoS Computational Biology. 2009;**5**(4): e1000363

[182] Domin G, Findeiß S, Wachsmuth M, Will S, Stadler PF, Mörl M. Applicability of a computational design approach for

synthetic riboswitches. Nucleic Acids Research. 2017;**45**(7):4108–4119

[183] Nakahira Y, Ogawa A, Asano H, Oyama T, Tozawa Y. Theophylline-dependent riboswitch as a novel genetic tool for strict regulation of protein expression in cyanobacterium Synechococcus elongatus PCC 7942. Plant and Cell Physiology. 2013;**54**(10):1724–1735

[184] Lynch SA, Gallivan JP. A flow cytometry-based screen for synthetic riboswitches. Nucleic Acids Research. 2009;**37**(1):184–192

[185] Topp S, Reynoso CM, Seeliger JC, Goldlust IS, Desai SK, Murat D, Shen A, Puri AW, Komeili A, Bertozzi CR, Scott JR. Synthetic riboswitches that induce gene expression in diverse bacterial species. Applied and Environmental Microbiology. 2010;**76**(23):7881–7884

[186] Ohbayashi R, Akai H, Yoshikawa H, Hess WR, Watanabe S. A tightly inducible riboswitch system in Synechocystis sp. PCC 6803. The Journal of General and Applied Microbiology. 2016;**62**(3):154–159

[187] Helliwell KE, Lawrence AD, Holzer A, Kudahl UJ, Sasso S, Kräutler B, Scanlan DJ, Warren MJ, Smith AG. Cyanobacteria and eukaryotic algae use different chemical variants of vitamin B12. Current Biology. 2016;**26**(8):999–1008

[188] Callura JM, Dwyer DJ, Isaacs FJ, Cantor CR, Collins JJ. Tracking, tuning, and terminating microbial physiology using synthetic riboregulators. Proceedings of the National Academy of Sciences. 2010;**107**(36):15898–15903

[189] Connor MR, Atsumi S. Synthetic biology guides biofuel production. BioMed Research International. 2010;**1**:541698

[190] Berens C, Suess B. Riboswitch engineering—making the all-important second and third steps. Current Opinion in Biotechnology. 2015;**31**:10–15

[191] Gay P, Le Coq D, Steinmetz M, Berkelman T, Kado CI. Positive selection procedure for entrapment of insertion sequence elements in gram-negative bacteria. Journal of Bacteriology. 1985;**164**(2):918–921

[192] Clerico EM, Ditty JL, Golden SS. Specialized techniques for site-directed mutagenesis in cyanobacteria. In: Rosato E, editor. Circadian Rhythms: Methods and Protocols. Vol. **362**. Totowa, NJ: Humana Press; 2007. p. 155–171

[193] Lea-Smith DJ, Vasudevan R, Howe CJ. Generation of marked and markerless mutants in model cyanobacterial species. Journal of Visualized Experiments. 2016;**111**:54001

[194] Saar KL, Bombelli P, Lea-Smith DJ, Call T, Aro EM, Müller T, Howe CJ, Knowles TP. Enhancing power density of biophotovoltaics by decoupling storage and power delivery. Nature Energy. 2018;**3**(1):75–81

[195] Dietsch M, Behle A, Westhoff P, Axmann IM. Metabolic engineering of Synechocystis sp. PCC 6803 for the photoproduction of the sesquiterpene valencene. Metabolic Engineering Communications. 2021;**13**:e00178

[196] Matsuoka M, Takahama K, Ogawa T. Gene replacement in

cyanobacteria mediated by a dominant streptomycin-sensitive rps12 gene that allows selection of mutants free from drug resistance markers. Microbiology. 2001;**147**(8):2077–2087

[197] Takahama K, Matsuoka M, Nagahama K, Ogawa T. High-frequency gene replacement in cyanobacteria using a heterologous rps12 gene. Plant and Cell Physiology. 2004;**45**(3): 333–339

[198] Viola S, Rühle T, Leister D. A single vector-based strategy for marker-less gene replacement in Synechocystis sp. PCC 6803. Microbial Cell Factories. 2014;**13**(1):4

[199] Zhang W, Song X, editors. Synthetic Biology of Cyanobacteria. Singapore: Springer; 2018. 1080 p

[200] Kunst F, Rapoport G. Salt stress is an environmental signal affecting degradative enzyme synthesis in Bacillus subtilis. Journal of Bacteriology. 1995;**177**(9):2403–2407

[201] Kojima K, Keta S, Uesaka K, Kato A, Takatani N, Ihara K, Omata T, Aichi M. A simple method for isolation and construction of markerless cyanobacterial mutants defective in acyl-acyl carrier protein synthetase. Applied Microbiology and Biotechnology. 2016;**100**(23): 10107–10113

[202] Pattharaprachayakul N, Lee M, Incharoensakdi A, Woo HM. Current understanding of the cyanobacterial CRISPR-Cas systems and development of the synthetic CRISPR-Cas systems for cyanobacteria. Enzyme and Microbial Technology. 2020;**140**:109619

[203] Cai F, Axen SD, Kerfeld CA. Evidence for the widespread distribution of CRISPR-Cas system in the Phylum Cyanobacteria. RNA Biology. 2013;**10**(5):687–693

[204] Hou S, Brenes-Álvarez M, Reimann V, Alkhnbashi OS, Backofen R, Muro-Pastor AM, Hess WR. CRISPR-Cas systems in multicellular cyanobacteria. RNA Biology. 2019;**16**(4):518–529

[205] Hsieh SC, Peters JE. Discovery and characterization of novel type ID CRISPR-guided transposons identified among diverse Tn7-like elements in cyanobacteria. Nucleic Acids Research. 2023;**51**(2):765–782

[206] Arévalo S, Pérez Rico D, Abarca D, Dijkhuizen LW, Sarasa-Buisan C, Lindblad P, Flores E, Nierzwicki-Bauer S, Schluepmann H. Genome engineering by RNA-guided transposition for Anabaena sp PCC 7120. ACS Synthetic Biology. 2024;**13** (3):901–912

[207] Wendt KE, Ungerer J, Cobb RE, Zhao H, Pakrasi HB. CRISPR/Cas9 mediated targeted mutagenesis of the fast-growing cyanobacterium Synechococcus elongatus UTEX 2973. Microbial Cell Factories. 2016;**15** (1):115

[208] Li H, Shen CR, Huang CH, Sung LY, Wu MY, Hu YC. CRISPR-Cas9 for the genome engineering of cyanobacteria and succinate production. Metabolic Engineering. 2016;**38**:293–302

[209] Zetsche B, Gootenberg JS, Abudayyeh OO, Slaymaker IM, Makarova KS, Essletzbichler P, Volz SE, Joung J, Van Der Oost J, Regev A, Koonin EV. Cpf1 is a single RNA-guided endonuclease of a class 2 CRISPR-Cas system. cell. 2015;**163**(3):759–771

[210] Ungerer J, Wendt KE, Hendry JI, Maranas CD, Pakrasi HB. Comparative genomics reveals the molecular determinants of rapid growth of the cyanobacterium Synechococcus elongatus UTEX 2973. Proceedings of the National Academy of Sciences. 2018;**115**(50):E11761–70

[211] Baldanta S, Guevara G, Navarro-Llorens JM. SEVA-Cpf1, a CRISPR-Cas12a vector for genome editing in cyanobacteria. Microbial Cell Factories. 2022;**21**(1):103

[212] Ziemann M, Reimann V, Liang Y, Shi Y, Ma H, Xie Y, Li H, Zhu T, Lu X, Hess WR. CvkR is a MerR-type transcriptional repressor of class 2 type VK CRISPR-associated transposase systems. Nature Communications. 2023;**14**(1):924

[213] Cengic I, Cañadas IC, Minton NP, Hudson EP. Inducible CRISPR/Cas9 allows for multiplexed and rapidly segregated single-target genome editing in Synechocystis sp. PCC 6803. ACS Synthetic Biology. 2022;**11**(9): 3100–3113

[214] Sengupta A, Liu D, Pakrasi HB. CRISPR-Cas mediated genome engineering of cyanobacteria. Methods in Enzymology. 2022;**676**:403–432

[215] Yao L, Cengic I, Anfelt J, Hudson EP. Multiple gene repression in cyanobacteria using CRISPRi. ACS Synthetic Biology. 2016;**5**(3):207–212

[216] Gordon GC, Korosh TC, Cameron JC, Markley AL, Begemann MB, Pfleger BF. CRISPR interference as a titratable, trans-acting regulatory tool for metabolic engineering in the cyanobacterium Synechococcus sp. strain PCC 7002. Metabolic Engineering. 2016;**38**:170–179

[217] Yao L, Shabestary K, Björk SM, Asplund-Samuelsson J, Joensson HN, Jahn M, Hudson EP. Pooled CRISPRi screening of the cyanobacterium Synechocystis sp PCC 6803 for enhanced industrial phenotypes. Nature Communications. 2020;**11**(1):1666

[218] Miao R, Jahn M, Shabestary K, Peltier G, Hudson EP. CRISPR interference screens reveal growth–robustness tradeoffs in Synechocystis sp. PCC 6803 across growth conditions. The Plant Cell. 2023;**35**(11):3937–3956

[219] Angermayr SA, Van der Woude AD, Correddu D, Vreugdenhil A, Verrone V, Hellingwerf KJ. Exploring metabolic engineering design principles for the photosynthetic production of lactic acid by Synechocystis sp. PCC6803. Biotechnology for Biofuels. 2014;**7**(1):99

[220] Yeremenko N, Jeanjean R, Prommeenate P, Krasikov V, Nixon PJ, Vermaas WF, Havaux M, Matthijs HC. Open reading frame ssr 2016 is required for antimycin A-sensitive photosystem I-driven cyclic electron flow in the cyanobacterium Synechocystis sp. PCC 6803. Plant and Cell Physiology. 2005;**46**(8):1433–1436

[221] Zhan J, Steglich C, Scholz I, Hess WR, Kirilovsky D. Inverse regulation of light harvesting and photoprotection is mediated by a 3′-end -derived sRNA in cyanobacteria. The Plant Cell. 2021;**33**(2):358–380

[222] Nawaz T, Gu L, Fahad S, Saud S, Harrison MT, Zhou R. Sustainable protein production through genetic engineering of cyanobacteria and use of atmospheric N2 gas. Food and Energy Security. 2024;**13**(2):e536

[223] Liu X, Tang K, Hu J. Application of cyanobacteria as chassis cells in synthetic biology. Microorganisms. 2024;**12**(7):1375

[224] Hachicha R, Elleuch F, Ben Hlima H, Dubessay P, de Baynast H, Delattre C, Pierre G, Hachicha R, Abdelkafi S, Michaud P, Fendri I. Biomolecules from microalgae and cyanobacteria: Applications and market survey. Applied Sciences. 2022;**12** (4):1924

[225] Govindasamy R, Gayathiri E, Sankar S, Venkidasamy B, Prakash P, Rekha K, Savaner V, Pari A, Thirumalaivasan N, Thiruvengadam M. Emerging trends of nanotechnology and genetic engineering in cyanobacteria to optimize production for future applications. Life. 2022;**12**(12):2013

[226] Angermueller C, Pärnamaa T, Parts L, Stegle O. Deep learning for computational biology. Molecular Systems Biology. 2016;**12**(7):878

[227] Zhang Q, Wang Y. AI in Biology: Transforming Genomic Research with Machine Learning. Computational Molecular Biology. 2024;**14**(3):106–114

[228] Seo E, Choi YN, Shin YR, Kim D, Lee JW. Design of synthetic promoters for cyanobacteria with generative deep-learning model. Nucleic Acids Research. 2023;**51**(13):7071–7082

[229] Pathania R, Srivastava A, Srivastava S, Shukla P. Metabolic systems biology and multi-omics of cyanobacteria: Perspectives and future directions. Bioresource Technology. 2022;**343**:126007

[230] Renganathan P, Gaysina LA, García Gutiérrez C, Rueda Puente EO, Sainz-Hernández JC. Harnessing engineered microbial consortia for xenobiotic bioremediation: Integrating multi-omics and AI for next-generation wastewater treatment. Journal of Xenobiotics. 2025;**15**(4):133

[231] Leão T, Wang M, Moss N, da Silva R, Sanders J, Nurk S, Gurevich A, Humphrey G, Reher R, Zhu Q, Belda-Ferre P. A multi-omics characterization of the natural product potential of tropical filamentous marine cyanobacteria. Marine Drugs. 2021;**19** (1):20

[232] Yue K, Zhu Y, Kai L. Cell-free protein synthesis: Chassis toward the minimal cell. Cells. 2019;**8**(4):315

[233] Li Z, Li S, Chen L, Sun T, Zhang W. Fast-growing cyanobacterial chassis for synthetic biology application. Critical Reviews in Biotechnology. 2024;**44**(3):414–428

[234] Novoveská L, Nielsen SL, Eroldoğan OT, Haznedaroglu BZ, Rinkevich B, Fazi S, Robbens J, Vasquez M, Einarsson H. Overview and challenges of large-scale cultivation of photosynthetic microalgae and cyanobacteria. Marine Drugs. 2023;**21** (8):445

Chapter 5

Assessing an Algae-Based Biorefinery for Wastewater Treatment and Pigment Extraction

Hana Faulds, Jing Li and Hamse Kjerstadius

Abstract

Rising populations are placing significant strain on traditional waste management, prompting the need for circular systems in urban projects to recycle valuable nutrients from wastewater. Introducing a biorefinery framework, microalgae cultivation on digestate effluents produces pigments for natural dyes, reducing nutrient release and offering eco-friendly alternatives to synthetic products. Successful extraction of phycocyanin (the blue pigment) was achieved with an optimal 75% wastewater (WW) and 20% media composition, yielding the highest phycobiliprotein content (132.93 mg/g DW (dry weight)) coupled with impressive nutrient removal rates of 7.41, 1.31, and 11.73 (mg.L - 1.d - 1) for NH^{4+}, $PO4^{3-}$, and TN (total nitrogen) respectively. The framework's scalability was evaluated in a Swedish urban development context, showcasing the potential of biorefinery concepts to convert waste into resources and supply natural pigments for textiles. Comparison with other studies showed that nutrient removal rates and PC yields are comparable to other wastewater-based systems with moderate biomass production, despite using a food-waste digestate stream. Our innovation also emphasized the high purity of phycocyanin (food-grade level), which many earlier studies don't report. Our study also provides insights into future development and discusses the limitations of practical applications.

Keywords: cyanobacteria, microalgae, natural dye, pigment, phycocyanin, spirulina, textiles, wastewater treatment

1. Introduction

For centuries, human consumption has followed a linear model, where resources are extracted, used only once, and discarded, typically relying on landfilling or incineration. The linear economy dominated after World War II until the concept of a circular economy gained traction in the late 2000s [1]. Wastewater has been increasingly seen as a resource, which provides added value derived from water, energy, nutrients, and materials [2, 3].

The concept of wastewater reuse, however, is not new. Ancient civilizations employed it for irrigation, fertilization, and disposal [4, 5]. Later, "sewage farms" applied wastewater to extensive land areas for both disposal and agricultural purposes [6], but increasing populations exposed their public health and environmental risks. The contemporary system was built to direct wastewater away from its source, collecting large volumes close to the point of environmental disposal. Modern centralized wastewater treatment plants (WWTPs) improved sanitation by removing waste from urban centers, yet they remain logistically complex, energy-intensive, and limited in nutrient recovery [7]. Although today's centralized WWTPs aim to improve sanitation, they are not yet optimal in terms of circularity. Reclaimed water requires storage and redistribution, while biosolids from mixed waste often lack safe reuse potential.

To overcome these inefficiencies, decentralized systems are emerging. One such example is RecoLab, a pioneering facility in Helsingborg, Sweden, which separates waste streams at the household level into black water (from the toilet), gray water (from baths, showers, and laundry), and food waste (from kitchen waste grinders) [8]. This source-separation model enables more targeted treatment and recovery, generating biogas and clean fertilizers while reclaiming high-quality water, paving the way for a circular urban metabolism.

Building on this principle, the present study explores a novel biorefinery concept that leverages microalgae to simultaneously treat nutrient-rich wastewater and produce high-value natural pigments. These pigments have potential applications in textiles and as sustainable alternatives to synthetic dyes in inks, paints, and coatings. The experimental work is part of the *Green Services and Value-Creating Water* project, a collaboration between RecoLab, Helsingborg City, and Lund University [9].

Using algae to overcome environmental problems as phytoremediation is a safe method, as macro-/microalgae are among the primary microorganisms that possess significant biosorption capacity to deactivate hazardous chemicals [10]. There is great potential shown for microalgae to complement other bioremediation processes to explore more effective industrial wastewater treatment for sulfur and metal recovery, as discussed in the role of sulfate-reducing bacteria (SRB) in the bioremediation of sulfur-rich wastewater [11]. While nutrient removal through microalgal treatment and the commercial use of algal pigments have been studied independently, a few approaches integrate both in a closed-loop biorefinery concept [12–14].

This study addresses that gap, focusing on the coupling of wastewater treatment with the coproduction of commercially relevant biochemicals, with specific attention to textile dye replacement [15]. Unlike earlier phases of the project, where mixed wastewater was used, this experiment employed food waste residuals, considered less contaminated and nutritionally sufficient for algal growth.

1.1 RecoLab treatment process

The black water and food waste streams have gained particular interest due to their elevated nutrient content. Food waste undergoes pasteurization at 70°C for one hour and then enters a digester, where anaerobic digestion produces biogas. The resulting digestates are nutrient-rich, especially in macronutrients like nitrogen and phosphorus. These elements are harnessed through struvite precipitation and nitrate extraction to produce biofertilizer from the liquid phase of the digestate. While this method effectively removes and recycles nutrients from wastewater, concerns over excessive fertilizer usage and nitrogen run-off, contributing to eutrophication, have prompted a reevaluation of EU regulations. This includes revisions under the EC Nitrates Directive (91/676/EEC) and nitrate vulnerable zone (NVZ) legislation. Restrictions on the amount of nitrogen that can be reintroduced to the land have been implemented, with similar restrictions on phosphate fertilizers currently being introduced [16].

1.2 Microalgae biorefinery concept

Recently, evoking global interest lies in an alternative valorization pathway: the cultivation of commercially valuable microalgae biomass using digestate effluents. The concept of using microalgae for nutrient remediation was initially put forth over 60 years ago by Oswald and Gotaas [17], pioneers in this field. Significant progress has been made afterward, with recent findings indicating between 85% and 100% removal efficiency by different microalgae species [18]. The microalgal system for removing nutrients has shown high removal efficiencies of up to 40 mg N L^{-1} d^{-1} and 4 mg P L^{-1} d^{-1} for nitrogen and phosphorus, respectively, with the added benefits of not requiring an organic carbon source and generating no sludge [19]. Despite microalgae's efficiency in removing both inorganic and organic contaminants, their use in wastewater treatment is still limited. In cases where they are implemented, it is often only a polishing step, and there is little emphasis on using the biomass produced for valuable products [20]. By coupling the process of waste management with microalgae production, this model not only addresses the challenges of nutrient remediation but also offers a solution to the high production costs associated with microalgae. This is achieved by replacing the use of expensive media and nutrient solutions with the digestate effluent stream. As the economic potential of microalgal biomass becomes perceivable, its utilization in wastewater treatment is likely to increase.

1.3 Valorization of microalgal biomass

Microalgae, versatile organisms rich in protein, vitamins, essential oils, antioxidants, pigments, and other valuable metabolites, hold significant biotechnological and environmental importance [21]. The main commercial applications for microalgae biomass include animal feed, nutraceuticals, cosmetics, and human nutrition. While much research on the biorefinery concept of microalgae grown on wastewater emphasizes the application of the resulting biomass as animal and fish feed, there is a growing interest in the extraction of pigments from microalgal biomass for use as natural dyes or ink solutions to replace petroleum-based products that dominate the market [22]. This study initiates a lab-scale experiment for growing algae on digestate effluents to extract valuable pigments.

1.4 Study objectives

This research aims to:

1. Evaluate the capacity of microalgae to remove nitrogen and phosphorus from synthetic wastewater.

2. Assess the quality and yield of natural pigments extracted from the cultivated biomass.

3. Examine the technical feasibility and circular economic potential of using this biomass in pigment-based applications.

1.5 Hypothesis

Microalgae cultivated in nutrient-rich wastewater can efficiently remove key pollutants while producing natural pigments suitable for use in the textile industry. This integrated process offers a scalable, decentralized solution for advancing circularity in wastewater management and sustainable production systems.

2. Materials and methods

This study comprised two main phases. Firstly, an initial growth experiment was conducted to test the photobioreactor system, assess the growth rate, and determine suitable media composition, followed by a main cultivation phase. All materials and methods refer to the master thesis project [23].

2.1 Materials

2.1.1 Arthrospira platensis (A. platensis) stock culture and inoculum preparation

Arthrospira platensis (A. platensis) starting culture was supplied by Health Algae. Upon its arrival, the 250 mL culture was placed under a white-fluorescent light to allow the culture to recover from any degradation during shipping. On the following day, the culture was divided into two 2 L Erlenmeyer flasks, each containing 1 L of Zarrouk's medium, and placed in a climate chamber (Pol-EKO Aparatura) [24] under a temperature of 25°C, a photoperiod of a 14:10 hours light: dark cycle, and a light intensity of 100%. On the third day, aeration tubes with gas stones were introduced for stirring, receiving 5,000 mL of air per minute from an aquarium air pump (IREENUO Q7 HQ-602). These cultures were maintained under the specified conditions, with 10% of the culture replaced with fresh media every 2–3 days until inoculation.

2.1.2 Media formulation

In this study, the Zarrouk medium used was modified based on the method outlined by Baker et al., 2021. It comprises three distinct solutions and was not autoclaved after mixing to avoid the precipitation of metals. Solutions 1 and 2 were autoclaved separately, and Solution 3 was sterilely filtered. After these steps, all three solutions were combined. The composition of the medium is as follows: 500 mL/L Solution 1 (16.8 g NaHCO3, 0.5 g K2HPO4 [per L]), 500 mL/L Solution 2 (2.5 g NaNO3, 1 g K2SO4, 1 g NaCl, 0.2 g MgSO4 · 7H2O, 0.04 g CaCl2, 0.01 g FeSO4 · 7H2O, 0.08 g NaEDTA [per L]), 1 mL/L Solution 3 (2.86 g H3BO3, 1.81 g MnCl2 · 4H2O, 0.222 g ZnSO4·7H2O, 0.079 g CuSO4 · 4H2O, 0.015 g MoO3 [per L]).

Wastewater samples were collected from the digested influent of the food-waste stream (which was kept separate from black and gray water streams) at the wastewater treatment facility, RecoLab. Upon its arrival at the treatment facility, food waste undergoes pasteurization at 70°C for one hour to eliminate any present pathogens, and then it enters the anaerobic digester (AD), where it is maintained at 37°C for several days to facilitate sedimentation. The liquid phase settled above the sediment was collected for use in this study. These collected samples were immediately autoclaved at 120°C for 30 minutes in accordance with lab safety protocols to eliminate any remaining microorganisms. Following autoclaving, the wastewater underwent filtration using standard Whatman filter paper to remove larger suspended particles, enhancing the liquid's transparency for improved light exposure. The samples were then stored at 4°C until use to prevent any biological activity and nutrient degradation [23].

Zarrouk medium and wastewater mixtures: different compositions of wastewater to medium were tested in duplicates: 100% FWW, 75% FWW + 25% Zarrouk, and 50% FWW + 50% Zarrouk, to compare the growth dynamics of A. platensis on increasing concentrations of wastewater. Phase I had a total working volume of 600 mL, and Phase II was scaled up to 4 L each [23].

2.1.3 Equipment and photobioreactor set-up

For both the growth experiment and the scale-up phase, a makeshift photobioreactor system was established. In the growth experiment, 1 L glass laboratory bottles (DWK Life Sciences DURAN) were arranged in sequence, featuring inlet and outlet tubing to facilitate aeration and air transfer. To minimize the risk of contamination, both the inlet and outlet were equipped with 0.22 μm filters. Each glass bottle was positioned equidistantly and sequentially from the light source, ensuring uniform light exposure without any shadowing effects between duplicates. During the scale-up phase, 5 L glass bottles were employed, maintaining the same lighting, tubing, and temperature systems [23].

2.2 Methods: experimental design

The first step of the experiment aimed to examine A. platensis cultivation using three different wastewater-based growth media: (1) 100% FWW, (2) 75% FWW supplemented with 25% Zarrouk medium, and (3) 50% FWW supplemented with 50% Zarrouk medium. The cultures, with a combined working volume of 600 mL, were incubated for 21 days under conditions identical to those used for the inoculum cultures. This experiment resulted in identifying the optimum ratio of wastewater to media. In the second step of the experiment, a composition of 75% FWW and 25% Zarrouk medium was exclusively used. The working volume for this experiment was scaled up to 4 L, and nutrient removal was monitored at increased intervals throughout the cultivation period. In both steps, the pH was adjusted to 9 using a 5 N NaOH solution, and it was maintained within the range of 9–11 during cultivation by using 5 N HCl or 5 N NaOH solutions. The light intensity was consistently set to 100%, following the climate chamber settings provided by white-fluorescent lights. The temperature was maintained at a constant 25 ± 3°C for the first experiment and at 28 ± 3°C for the second; however, humidity was not controlled. All experiments were performed in duplicate under nonaseptic conditions, with continuous aeration provided by an air pump at a rate of 5 L air/min. The experimental design and climate chamber conditions are visually outlined in **Figure 1**. Replicates of samples were used for experiments [23].

2.3 Analysis

2.3.1 pH and conductivity

During both experiments, pH and conductivity were monitored using the HANNA Combo® HI98129 instrument. This instrument was also used to check the pH of wastewater samples and other growth media used.

2.3.2 Dry weight, optical density, and standard curve

To facilitate sampling, a calibration curve was created to examine the relationship between dry weight and optical density. A stock culture of A. platensis and five serial dilutions, each with a 20 mL volume, were prepared. The optical density for each dilution was then measured using a spectrophotometer (Ultrospec 2100pro, Amersham Biosciences) at a wavelength of 565 nm. Dry weight determination followed a standard protocol: 0.45 μm membrane filters were microwaved for 4 minutes, placed in a desiccator overnight, and their initial weight was recorded. Afterward, 10 mL of each dilution was filtered, washed three times with 5* × 10 mL of cold distilled water to collect the algal cells, microwaved for 8 minutes, and left to dry in the desiccator overnight. The final weight of the filters was recorded, and the dry weight was calculated using the equation below [23].

$$Dryweight(mg/mL) = \frac{initialweightoffilter - finalweightoffilter}{volumefiltered(mL)} \quad (1)$$

A standard curve was plotted in **Figure 2** for the data within the range of 0–1.4 g/L dry weight of culture, and a linear regression was performed to obtain the linear equation and the R-squared value of the curve.

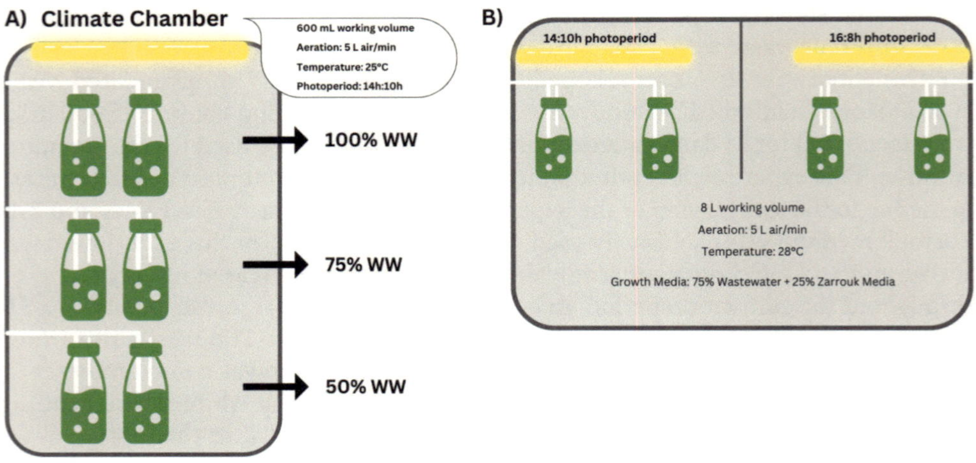

Figure 1.
Graphic layout and parameter specifications for (A) optimization experiment 1 and (B) scale-up experiment 2.

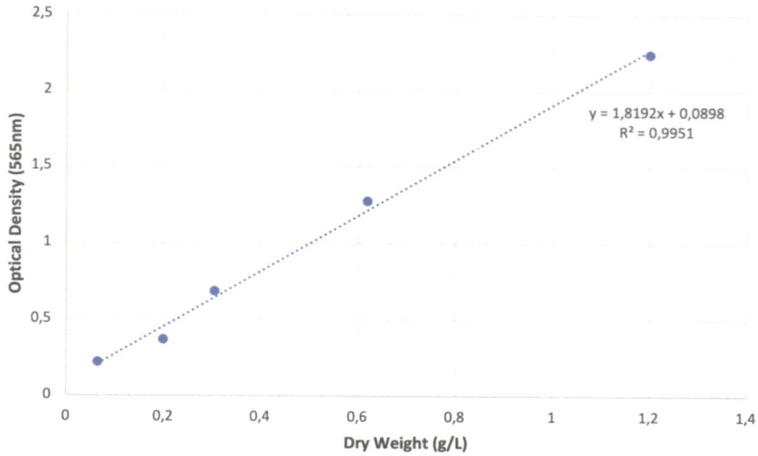

Figure 2.
Linear correlation between optical density at 565 nm and dry weight measurements (g/L).

2.3.3 Phycobiliprotein content

Phycobiliproteins (PBPs) are a special class of water-soluble pigment–proteins found in cyanobacteria (blue-green algae), red algae, cryptophytes, and some other microalgae. There are several types of PBPs, which are named for their absorption color:

- Phycocyanin (PC): Blue pigment
- Allophycocyanin (APC): Bluish pigment
- Phycoerythrin (PE): Red pigment
- Phycoerythrocyanin (PEC): bluish-red (pink-orange)

The quantification of PBPs in the biomass produced during both experiments followed similar methods outlined by Arashiro et al. [25], with relevant adaptations. Briefly, 5 mL samples, taken every 2–4 days, were immediately centrifuged at 1,160 g for 15 minutes, washed twice with cold distilled water, and the resulting biomass pellet was frozen at - 21°C for further quantification. The pellets were then resuspended in 4 mL of concentrated 37% HCl and kept in the dark at 4°C for 24 hours. Optical densities were then measured at 280 nm, 562 nm, 615 nm, 652 nm, and 620 nm. The PC, APC, and PE concentrations and yield were calculated using the equations proposed by [26] as listed below.

$$PC(mg/mL) = [A_{615} - (0.474 * A_{652})]/5.34 \tag{2}$$

$$APC(mg/mL) = [A_{652} - (0.208 * A_{615})]/5.09 \tag{3}$$

$$PE(mg/mL) = [A_{562} - (2.41 * PC) - (0.849 * APC]/9.62 \tag{4}$$

Purity of PC was determined according to the ratio A620/A280, as defined by Cuellar-Bermudez et al. (2015), where A615, A652, A280, A620, and A562 correspond to the absorbance measured at the respective wavelengths [27]. The purity of PC is related to the total amount of protein present in the extraction product [28].

2.3.4 Nutrient removal

Nutrient removal was measured by monitoring concentrations of ammonium (NH4–N), phosphorus ($PO4^{3-}$ -P), total nitrogen (TN), and total organic carbon (TOC) using the following Hach test kits: LCK303 ammonium (2–47 mg/L), LCK302 ammonium (47–130 mg/L), LCK350 phosphorus Total (2–20 mg/L), LCK348 phosphorus total (0.5–5 mg/L), LCK438 LATON TN (100–250 mg/L), and LCK386 Total Organic Carbon (30–300 mg/L), following the test protocols provided using a portable spectrophotometer (Hach DR3900).

2.3.5 Calculations

For simplicity, the specific growth rate (d – 1) of A. platensis throughout this project was estimated using the equation below, where X is cell concentration (g/L) and t is time (days).

$$X(t) = X_0 * e^{\mu t} \tag{5}$$

Biomass productivity (mg · L^{-1} – 1 · d – 1) was determined by Eq. 6, where Xmax (g · L – 1) was the maximal biomass concentration, X0 (g · L – 1) was the initial biomass concentration, and Tmax was the time for biomass to reach maximum (days).

$$Biomassproductivity\left(\text{mg.L}^{-1}.\text{d}^{-1}\right) = \left[X_{max} - \frac{X_0}{T_{max}}\right] * 1000 \tag{6}$$

The removal efficiency (%) of pollutants from wastewater was evaluated by Eq. 7, where Cin (mg · L – 1) was the initial concentration of the pollutant (NH4 PO_4^{3-}, and TN), and Cout (mg · L – 1) was the concentration of the residual pollutant when the biomass concentration reached its maximum.

$$Removalefficiency(\%) = \frac{C_{in} - C_{out}}{C_{in}} x100 \tag{7}$$

The average removal rate (mg·L – 1·d – 1) of each pollutant was calculated by Eq. 8, where T_{max} was the time (days) for biomass to reach its maximum.

$$Av.Removalrate\left(mg.L^{-1}.d^{-1}\right) = \frac{C_{in} - C_{out}}{T_{max}} \tag{8}$$

2.3.6 Statistical analysis

All assays were performed in duplicates, and statistical analysis was conducted using JASP (JASP-stats, Amsterdam, Netherlands). The average values measured during both experiments were analyzed using multifactor analysis of variance (ANOVA), and the significant differences among growth media variations and PC production were determined using a two-tailed T-test at a 95% confidence interval.

3. Results

3.1 Growth analysis

The initial growth experiment of A. platensis involved different concentrations of digestate effluent, as shown in **Figure 3A**. The results for cultivation grown in 100% WW indicate minimal growth. On the contrary, cultivations on 75% and 50% WW exhibited notable biomass concentrations, reaching 1.6 g L − 1 after 17 days and 2.09 g L − 1 after 18 days, respectively.

The absence of growth with 100% WW indicates that wastewater alone may lack essential elements such as alkalinity, salinity, and additional nutrient components like potassium and trace elements, which are necessary for the optimal growth and health of A. platensis. Potential factors, such as ammonium-inhibition or the influence of specific bacterial populations, call for more in-depth investigation. Given that the objective was to establish growth primarily on WW while ensuring

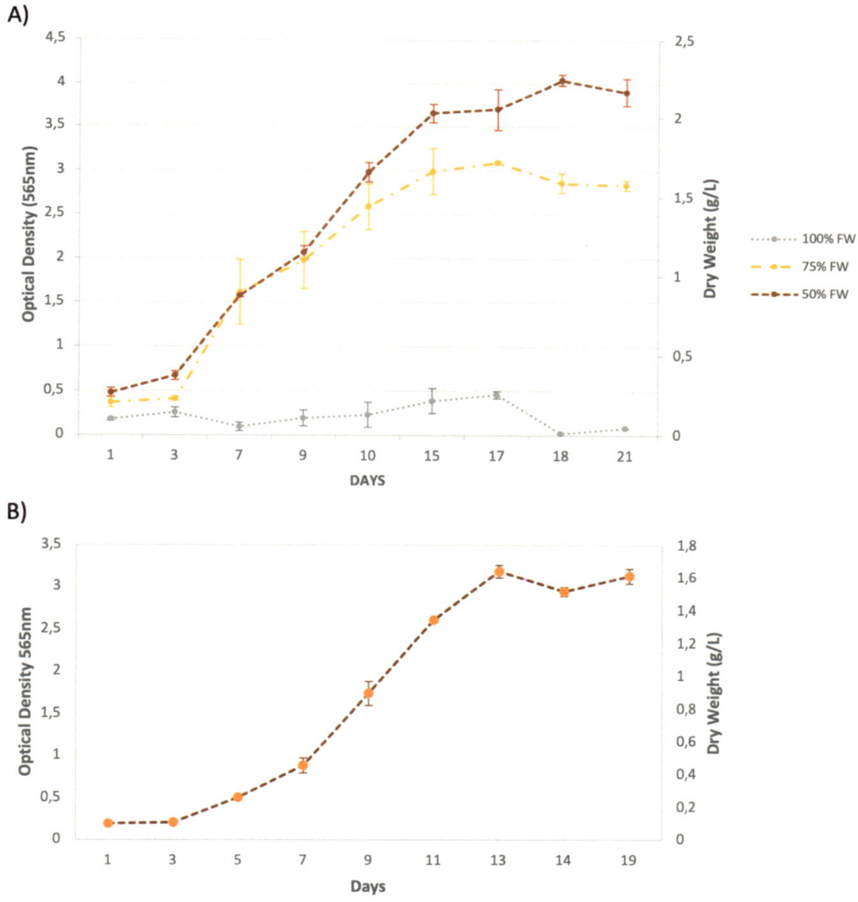

Figure 3.
Biomass growth of A. platensis during (A) the optimization experiment (working volume: 600 mL) on growth media containing different amounts of WW (100%, 75%, 50%) supplemented with Zarrouk media, and (B) main cultivation (working volume: 4 L) in 75% WW, measured as optical density at 565 nm and converted to dry weight (g/l).

substantial biomass growth, the growth rate measured in the 75% WW cohort was considered sufficient and was thus selected as the optimal medium composition for the main cultivation, despite the higher biomass concentration observed in the 50% WW cohort. During the main cultivation, the total volume increased from 600 mL to 4 L, achieving a peak biomass concentration of 1.64 g·L – 1 at a growth rate of 0.34 g·d – 1 after only 13 days, as shown in **Figure 3B** [23].

3.2 Nutrient removal

Ammonium (NH^{4+}), orthophosphate ($PO4^{3-}$) and TN, illustrated in **Figure 4**, functioned as nutrient markers. Their concentrations were monitored at specific time points during the cultivation of A. platensis to evaluate the efficiency and rate of nutrient removal.

Results in **Figure 4** show that the removal rate and efficiency were impressive for all three nutrient markers, averaging 7.41, 1.31, and 11.73 (mg·L – 1.d – 1) for NH^{4+}, $PO4^{3-}$, and TN, respectively. The initial concentrations of nutrients (NH^{4+}, $PO4^{3-}$, and TN) measured before A. platensis inoculation were 183.55, 20.6, and 96.6 (mg/L), respectively. **Figure 3** also demonstrates that, in every instance, concentrations fell below the accurate detection range by day 10 of cultivation, and both nitrogen and phosphorus sources were approaching depletion by day 14. This rapid nutrient removal aligns with the observed lack of biomass growth after day 14, indicating that the cultivations entered a stationary phase. Efficiency of removal (%) was calculated at 96%, 95%, and 88% [23].

3.3 Phycocyanin concentration and purity

The PBP content of the biomass was examined during both growth analyses, as it is recognized to have a positive correlation with cyanobacterial biomass concentration [29]. **Figure 5** illustrates the concentration and purity of PC extracted from the final biomass. Concentration indicates the amount of PC per gram of dry

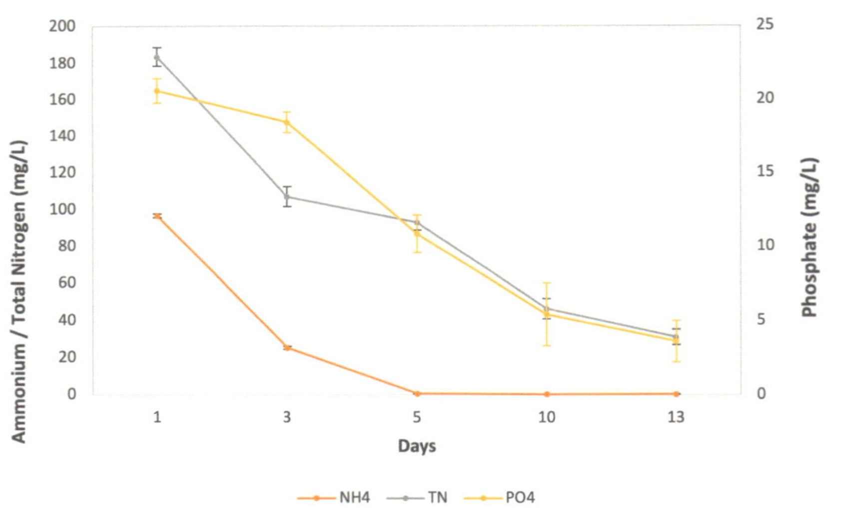

Figure 4.
Nutrient removal during main cultivation (working Vol. 4 L) as shown by ammonium ($NH4^+$), total nitrogen (TN), and phosphate ($PO4^{3-}$) consumption by A. platensis throughout cultivation.

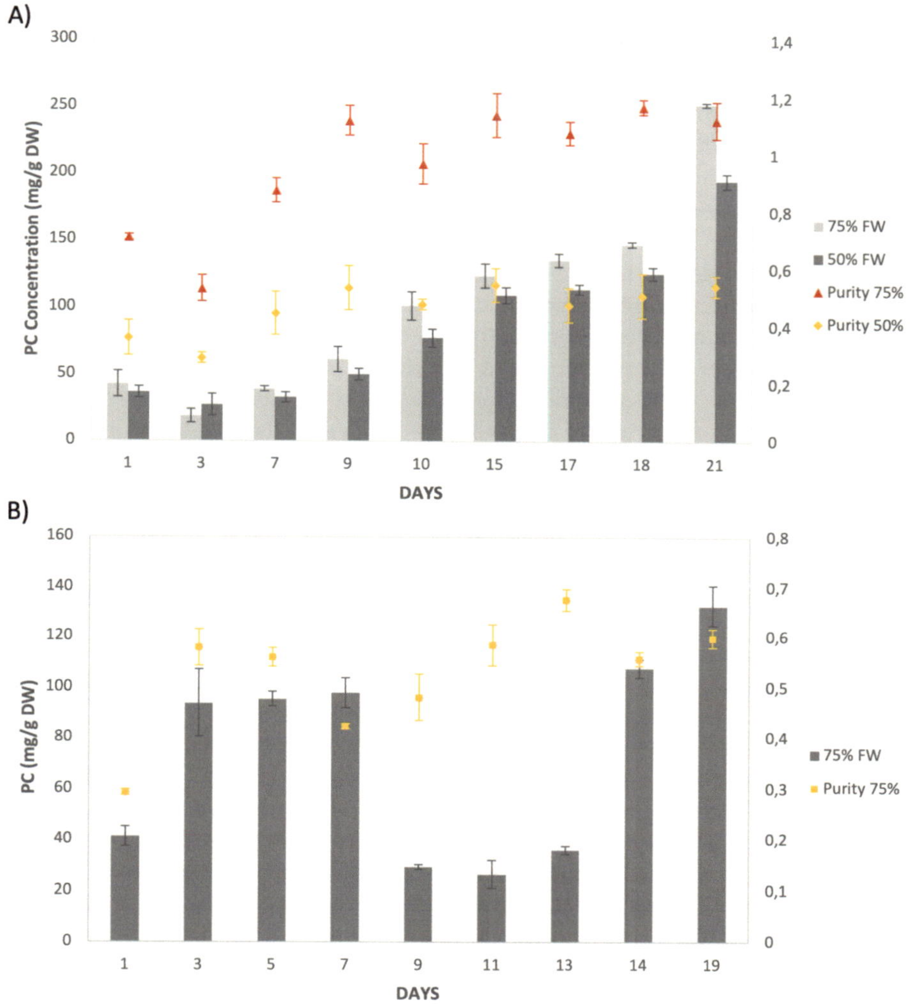

Figure 5.
Phycocyanin content in total biomass and respective purity during (A) Phase I of A. platensis cultivation at 600 mL working volume with different concentrations of WW and (B) Phase II at a working volume of 4 L with 75% WW concentration in the growth media [23].

weight, while purity defines the relationship between the presence of PC and other contaminating proteins in the extracted sample [30].

The purity ratio, determined by the A_{620}/A_{280} ratio, indicates high purity when values are above or equal to 0.7, signifying food-grade pigment, and values above or equal to 4.0 correspond to reagent and analytical-grade pigment [31]. In the initial growth analysis (**Figure 4B**), there was a progressive increase in PC content measured throughout cultivation, reaching final concentrations of 252.76 mg/g DW and 195.2 mg/g DW for the 50% and 75% WW cultivations, respectively. Previous studies have observed similar patterns in the influence of wastewater-based media composition on pigment concentration, with higher amounts of NaHCO3 and NaCl resulting in the maximum PC concentration [29]. Arashiro et al. (2020) also reported similar findings, achieving maximum PC concentrations (approx. 200 mg/g DW) with a combination of standard growth medium and wastewater [14]. Factors such as nitrogen concentration, bacterial populations, salinity, and pH can impact PBP concentration during cultivation, suggesting a need for further analysis of these variables. In summary, the optimal conditions for PBP production were found with a media composition of 75% wastewater and 25% Zarrouk media [23].

3.4 Assessing scale-up potential

Although the initial lab-scale testing was promising, scaling up any algae cultivation plant presents challenges and requires a comprehensive commercial assessment. However, a full economic feasibility study was beyond this project's scope. Typically, production costs decrease with increased capacity. However, the proposed model is constrained by the volume of liquid effluent leaving the anaerobic digester (AD), determined by the total incoming volume from residential usage. Due to innovative technology and water-saving ambitions in the Oceanhamnen development, an average of only 5 L per person per day of wastewater enters the facility, compared to 250 L per person per day at a traditional WWTP. Approximately 4.5 L is estimated to leave the AD as liquid effluent, utilized as microalgae growth media. Considering these volumes and assuming a total initial population of 1,000 residents, as a conservative estimation, the process feasibility was assessed with an extraction capacity between 5,625 and 6,000 L/day. Assuming a constant growth rate (0.34 d – 1) and maximum biomass concentration (1.64 g/L) from lab-scale growth analysis, the estimated annual biomass production is 2,398.5–2,558.4 kg, equating to a minimum annual extraction of 191.88–204.6 kg (8% of total biomass) C-phycocyanin. To consider this process for commercial application, the system would need to have an annual extraction of closer to one ton, around 10 times the current estimated extraction capacity. The mass balance for nitrogen and phosphate over the proposed cultivation system has been outlined in **Figure 6** below, illustrating the potential use of microalgae as a nutrient extraction step during wastewater treatment, where most of the nitrogen and phosphate are removed from the water stream by the resulting biomass [23].

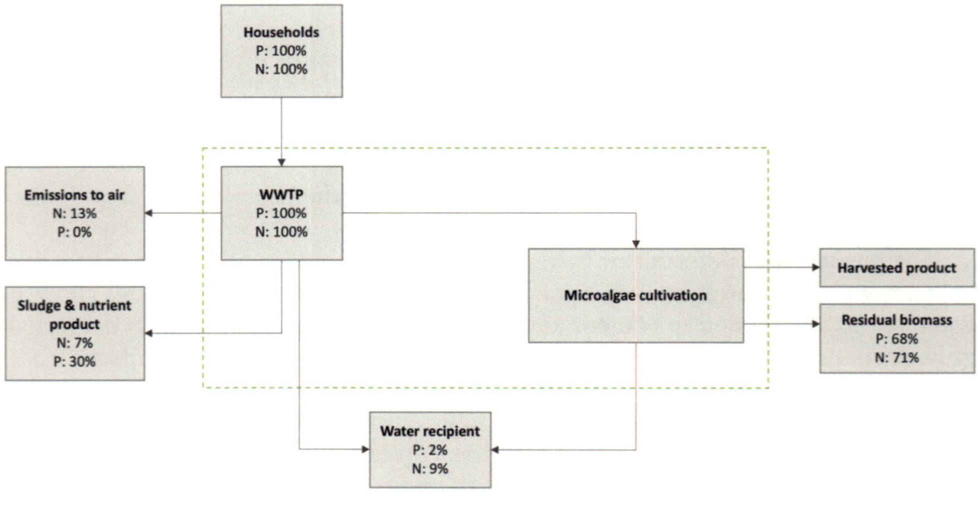

Figure 6.
Mass balance for nitrogen (N) and phosphate (P) over the proposed cultivation system[23].

4. Discussion

This study has demonstrated the dual potential of A. platensis cultivation on food-waste-derived wastewater, achieving high nutrient removal efficiencies while producing a commercially relevant pigment: PC.

4.1 Growth dynamics

The observed stationary phase around day 14 suggests that, beyond nutrient depletion, additional factors may have limited productivity. Possible explanations include light penetration constraints in denser cultures, pH drift due to metabolic activity, and accumulation of inhibitory metabolites. Similar growth slowdowns in wastewater-based media were reported by Arashiro et al. [14] and Papadopoulos et al. [29], supporting the need for optimized photoperiods, mixing, and CO_2 supplementation.

4.2 Nutrient removal

Our nutrient removal results align with previous studies [14]. Their reported values, ranging from 52% to 100% for TN and between 73% and 82% for phosphate removal during cultivation on industrial wastewater, showed no significant effect from varied light conditions. The particularly high ammonium removal rate (7.41 mg · L − 1 · d − 1) indicates that digestate-derived wastewater provides a suitable nitrogen source, although ammonium toxicity may partly explain poor growth in 100% WW treatments. This reinforces the importance of media balancing when targeting both remediation and biomass productivity.

Observing the removal of COD, heavy metals, and additional forms of nitrogen throughout cultivation could offer a more comprehensive evaluation of this strain's capability to treat the wastewater stream. This aspect is recommended for future research.

4.3 Pigment production

PC yields (195–253 mg/g DW) are comparable to those reported in mixed wastewater-media systems [14] [29], as shown in **Table 1**. Importantly, the purity values achieved (A620/A280 ≥0.7) reached food-grade standards, indicating strong potential for textile and nutraceutical applications. The method employed here, while effective, could benefit from downstream process optimization to increase recovery rates and reduce costs, as highlighted by Borowitzka [31].

4.4 Comparison with previous studies

The nutrient removal rates and PC yields are comparable to other wastewater-based systems, despite using a food-waste digestate stream (novel source); see **Table 1**.

Our innovation emphasized the high purity of PC (food-grade level), which many earlier studies don't report. It is also noted that the biomass productivity is moderate compared to some, but still promising given decentralized urban wastewater volumes.

Study	Microalgae species	Wastewater type and medium	Biomass productivity ($g \cdot L^{-1}$)	N removal (%)	P removal (%)	Phycocyanin yield (mg/g DW)	Notes
This study (2025)	*Arthrospira platensis*	75% food-waste digestate + 25% Zarrouk	1.64	96% (NH_4^+), 88% (TN)	95%	195–253	High purity (≥0.7 A620/A280); poor growth in 100% WW
Cañizares-Villanueva et al. [18]	Cyanobacteria	Diluted swine wastewater	1.2–1.5	~80–90%	~70%	~180	Aerated systems: strong dependency on dilution ratio
Zhu [19]	Mixed microalgae	Synthetic wastewater	1.0–2.0	85–100%	80–90%	n/a	Review – emphasizes the need for nutrient balancing
Arashiro et al. [14]	Cyanobacteria	Industrial wastewater + growth medium	1.8–2.2	52–100%	73–82%	~200	Comparable pigment yields; light not limited.
Papadopoulos et al. [29]	*Arthrospira platensis*	Brewery wastewater	1.5–1.8	~90%	~80%	~220	Higher PC with elevated $NaHCO_3$ and NaCl.
Borowitzka [31]	Review (various)	Various	0.8–2.5	70–95%	60–85%	100–400	Notes commercial potential but stresses cost barriers

Table 1.
Comparison with other similar studies.

4.5 Scale-up potential

Borowitzka [31] briefly reviews the main existing and potential high-value products that can be derived from microalgae and considers their commercial development, with a particular focus on the various aspects that need to be addressed on the path to commercialization, using the experience gained in the commercialization of existing algae products. While our mass balance indicates feasibility at the pilot scale, the production volume estimated (~200 kg C-phycocyanin/year) is well below industrial demand thresholds (>1 ton/year). Similar economic challenges have been noted in other microalgae biorefineries [19, 31] for diverse applications. Scaling may require integration with multiple urban digesters, improved photobioreactor efficiency, or coupling with additional high-value coproducts (e.g., antioxidants, bioactive peptides) to strengthen economic viability. When integrating wastewater treatment processes for sustainable energy recovery, microalgae have shown impressive energy output potential, with algal biomass having an energy potential of 60–160 MWh per hectare per year, depending on the seasonality and treatment method [32]. Resource recovery overall is still limited in practice, as noted in a critical review, which states that despite many technical options, "only a few technologies have been implemented and a shift from wastewater treatment plants towards water resource recovery facilities still seems far away" [33]. Future endeavors need to be undertaken to advance the role of microalgae in resource recovery further.

4.6 Future work

This study focused on nutrient removal and pigment extraction but did not assess COD, heavy metals, or microbiome shifts during cultivation. Future work should include:

- Expanded monitoring of additional pollutants (COD, metals)
- Investigation of microbial community dynamics and their role in productivity
- Optimization of light intensity, photoperiod, and CO_2 aeration
- A techno-economic assessment of scale-up under decentralized wastewater treatment models

To investigate algae's role in wastewater treatment, there is huge potential that can be further explored for a circular future; for example, soil enzymes from microalgae have great potential for their application in biotechnology and agriculture [34].

5. Conclusion(s)

This paper presents a biorefinery framework for cultivating microalgae using digestate effluent and extracting valuable pigments from the biomass. Successful growth of A. platensis on municipal waste streams confirms its potential. While wastewater alone didn't sustain growth, 75% and 50% wastewater cultivations showed sufficient biomass growth. The optimal 75% wastewater and 20% media composition produced the highest PBP content. Impressive nutrient removal rates were observed for nitrogen (N) and phosphorus (P) in all cultivations, indicating nutrient recovery efficiency. We assessed scalability for the entire Oceanhamnen development, a step toward circular community development.

Despite these achievements, further optimization is needed to improve growth rate and biomass productivity, exploring factors like photoperiod, light intensity, and CO_2 levels. Comprehensive evaluation, including COD, heavy metals, and additional nitrogen forms removal, is essential. Future work should address the bacterial community's impact on biomass productivity and PC concentration and purity. This framework holds promise for the textile industry, paints, inks, and natural dye products but requires ongoing refinement for broader applications and enhanced efficiency. Microalgae, as a resource recovery medium in general, is a great candidate to be further investigated.

Acknowledgments

This project was supported by Kompetensfonden within the project GRÖNA TJÄNSTER OCH VÄRDESKAPANDE VATTEN: Cirkulerande system för odling av mikroalger och rening av källsorterad VA på RecoLab i Helsingborg (GREEN SERVICES AND VALUE CREATING WATER: Circulating system for cultivation of microalgae and purification of source-sorted VA at RecoLab in Helsingborg), with project number 2109, project stage 2. Appreciation to project partners\support, RECOLAB, for providing materials, water samples, and onsite learning and knowledge sharing, and the Department of Service Studies for project management and inspiration. The content of this project is mainly contributed by the Master Thesis project by Hana Faulds at Lund University.

Conflict of interest

The authors declare no conflict of interest.

Notes/thanks/other declarations

Place any other declarations, such as "Notes", "Thanks", etc., before the References section. Assign the appropriate heading.

Author details

Hana Faulds[1,2], Jing Li[1*] and Hamse Kjerstadius[3]

1 Division of Water Resources Engineering, Lund University, Lund, Sweden

2 OneBio Innovation Centre, Cape Town, South Africa

3 Nordvästra Skånes Vatten och Avlopp AB (NSVA), Helsingborg, Sweden

*Address all correspondence to: jing.li@tvrl.lth.se

References

[1] Aggeri, Franck, "From waste to urban mines: A historical perspective on the circular economy," No. Special Issue 23, Pp. 10–13, 2021 November. [Online]. Available from: http://journals.openedition.org/factsreports/6530 [Accessed: 2025-August-20]

[2] Zacharof MP, Lovitt RW. Adding value to wastewater by resource recovery and reformulation as growth media: Current prospects and potential. Journal of Water Reuse and Desalination. 2015;**5**(4):473–479. DOI: 10.2166/WRD.2015.117

[3] Wang C, Deng SH, You N, Bai Y, Jin P, Han J. Pathways of wastewater treatment for resource recovery and energy minimization towards carbon neutrality and circular economy: Technological opinions. Frontiers in Environmental Chemistry. 2023;**4**(October):1–7. DOI: 10.3389/fenvc.2023.1255092

[4] Soares A. Wastewater treatment in 2050: Challenges ahead and future vision in a European context. Environmental Science and Ecotechnology. 2020;**2**:100030. DOI: 10.1016/J.ESE.2020.100030

[5] deFeo G, et al. The Historical Development of Sewers Worldwide. Sustainability. 2014;**6**(6):3936–3974. DOI: 10.3390/SU6063936

[6] Angelakis AN, Snyder SA. Wastewater treatment and reuse: Past, present, and future. Water. 2015;**7**(9):4887–4895. DOI: 10.3390/W7094887

[7] Zaharia C. Decentralized wastewater treatment systems: Efficiency and its estimated impact against onsite natural water pollution status. A Romanian case study. Process Safety and Environmental Protection. 2017;**108**:74–88. DOI: 10.1016/J.PSEP.2017.02.004

[8] "RecoLab". [Online]. Available from: https://www.recolab.se/ [Accessed: 2025-August-23]

[9] Andersson M, Fredriksson A, Cecilia HAUX, KJERSTADIUS H, LI J, Persson KM. Gröna Tjänster Och Värdeskapande Vatten En Rapport Från En Fallstudie Av Innovativ. Lund: Media-Tryck: 2023. p. 30. Available from: https://lup.lub.lu.se/search/files/159196335/Rapport_Malin_tillg.pdf

[10] Shahi Khalaf Ansar B, et al. Removal of organic and inorganic contaminants from the air, soil, and water by algae. Environmental Science and Pollution Research. 2022;**30**(55):116538–116566. DOI: 10.1007/S11356-022-21283-X

[11] Bagheri Novair S, Biglari Quchan Atigh Z, Asgari Lajayer B, Shu W, Price GW. The role of sulphate-reducing bacteria (SRB) in bioremediation of sulphate-rich wastewater: Focus on the source of electron donors. Process Safety and Environmental Protection. 2024;**184**:190–207. DOI: 10.1016/J.PSEP.2024.01.103

[12] Al-Jabri H, Das P, Khan S, Thaher M, Abdulquadir M. Treatment of Wastewaters by Microalgae and the potential applications of the produced biomass—A Review. Water. 2020;**13**(1):27. DOI: 10.3390/W13010027

[13] Nguyen LN, et al. Nutrient removal by algae-based wastewater treatment. Current Pollution Reports. 2022;**8**(4):369–383. DOI: 10.1007/s40726-022-00230-x

[14] Arashiro LT, Boto-Ordóñez M, Van Hulle SWH, Ferrer I, Garfí M, Rousseau DPL. Natural pigments from microalgae grown in industrial wastewater. Bioresource Technology. 2020;**303**:122894. DOI: 10.1016/J.BIORTECH.2020.122894

[15] Gomes M, Felgueiras HP, Leite BR, Soares GMB. Colourful protection: Challenges and perspectives of antibacterial pigments extracted from bacteria for textile applications. Antibiotics. 2025;**14**(5):520. DOI: 10.3390/ANTIBIOTICS14050520

[16] Abinandan S, Subashchandrabose SR, Venkateswarlu K, Megharaj M. Nutrient removal and biomass production: Advances in microalgal biotechnology for wastewater treatment. Critical Reviews in Biotechnology. 2018;**38**(8):1244–1260. DOI: 10.1080/07388551.2018.1472066

[17] Oswald WJ, Gotaas HB. Photosynthesis in Sewage Treatment. Transactions of the American Society of Civil Engineers. 1957;**122**(1):73–97. DOI: 10.1061/TACEAT.0007483

[18] Cañizares-Villanueva RO, Domínguez AR, Cruz MS, Ríos-Leal E. Chemical composition of cyanobacteria grown in diluted, aerated swine wastewater. Bioresource Technology. 1995;**51**(2–3):111–116. DOI: 10.1016/0960-8524(94)00099-M

[19] Zhu L. Microalgal culture strategies for biofuel production: A review. Biofuels, Bioproducts and Biorefining. 2015;**9**(6):801–814. DOI: 10.1002/BBB.1576

[20] Arashiro LT, Boto-Ordóñez M, Van Hulle SWH, Ferrer I, Garfí M, Rousseau DPL. Natural pigments from microalgae grown in industrial wastewater. Bioresource Technology. 2020;**303**:122894. DOI: 10.1016/j.biortech.2020.122894

[21] Dolganyuk V, et al. Microalgae: A promising source of valuable bioproducts. Biomolecules. 2020;**10**(8):1–24. DOI: 10.3390/BIOM10081153,

[22] Geng Y, et al. Microalgal biorefineries: A systematic review of technological trade-offs and innovation pathways. Biotechnology for Biofuels and Bioproducts. 2025;**18**(1):1–20. DOI: 10.1186/S13068-025-02694-7

[23] Faulds H, "Production of natural pigments from A. platensis grown on wastewater from local treatment facility.," 2023. [Online]. Available from: https://lup.lub.lu.se/student-papers/record/9118915 [Accessed: 2025-September-2]

[24] "POL-EKO sp.k. - Information and distributors around the world - Directindustry." [Online]. Available from: https://www.directindustry.com/soc/pol-eko-spk-104809.html [Accessed: 2025-August-23]

[25] Arashiro LT, Boto-Ordóñez M, Van Hulle SWH, Ferrer I, Garfí M, Rousseau DPL. Natural pigments from microalgae grown in industrial wastewater. Bioresource Technology. 2020;**303**:122894. DOI: 10.1016/J.BIORTECH.2020.122894

[26] Bennett A, Bogobad L. Complementary chromatic adaptation in a filamentous blue-green Alga.

Journal of Cell Biology. 1973;**58**(2): 419–435. DOI: 10.1083/JCB.58.2.419,

[27] Cuellar-Bermudez SP, Aguilar-Hernandez I, Cardenas-Chavez DL, Ornelas-Soto N, Romero-Ogawa MA, Parra-Saldivar R. Extraction and purification of high-value metabolites from microalgae: Essential lipids, astaxanthin and phycobiliproteins. Microbial Biotechnology. 2015;**8**(2):190–209. DOI: 10.1111/1751-7915.12167

[28] Boussiba S, Richmond AE. Isolation and characterization of phycocyanins from the blue-green alga Spirulina platensis. Archives of Microbiology. 1979;**120**(2):155–159. DOI: 10.1007/BF00409102

[29] Papadopoulos KP, et al. Cultivation of Arthrospira platensis in Brewery Wastewater. Water. 2022;**14**(10):1547. DOI: 10.3390/W14101547

[30] Ryckebosch E, et al. Influence of extraction solvent system on the extractability of lipid components from the biomass of Nannochloropsis gaditana. Journal of Applied Phycology. 2014;**26**(3):1501–1510. DOI: 10.1007/s10811-013-0189-y

[31] Borowitzka MA. High-value products from microalgae—their development and commercialisation. Journal of Applied Phycology. 2013;**25**(3):743–756. DOI: 10.1007/S10811-013-9983-9

[32] "Renewable energy from wastewater grown microalgae - a concept for nutrient recycling and sustainable energy recovery | Lund University". [Online]. Available from: https://www.lunduniversity.lu.se/lup/publication/5472626 [Accessed: 2025-August-25]

[33] Kehrein P, Van Loosdrecht M, Osseweijer P, Garfí M, Dewulf J, Posada J. A critical review of resource recovery from municipal wastewater treatment plants – Market supply potentials, technologies and bottlenecks. Environmental Science (Cambridge). 2020;**6**(4):877–910. DOI: 10.1039/C9EW00905A

[34] Shoarnaghavi MA, et al. Soil algae enzymes and their biotechnological applications. Systems Microbiology and Biomanufacturing. 2022;**2**(4):589–606. DOI: 10.1007/S43393-022-00095-7

Chapter 6

Limnospira platensis LEGE 231238: Exploring Commercial Potential through Comprehensive Biochemical and Taxonomic Analysis

Guilherme Scotta Hentschke, Flavio Oliveira, Graciliana Lopes, Tiago Conde, Tánia Melo, Marisa Pinho, Maria do Rosário Domingues, Hugo Pereira, Sara Ferreira, Claudia Hoepfner, Vitor Vasconcelos and Daniel Guzmán

Abstract

The strain *Limnospira platensis* LEGE 231238, isolated from the saline Alalay Lake in Bolivia, was identified through a polyphasic approach. The biochemical characterization revealed that the dry biomass presented (% of dry weight) protein (45.40 ± 0.54), carbohydrates (27.84 ± 0.25), ash (19.90 ± 0.32), lipids (6.86 ± 0.18), and C:N ratio (5.54 ± 0.17). Lipid extracts revealed 46.23 ± 0.60% polyunsaturated fatty acids (FA), 41.24 ± 0.72% saturated FA, with palmitic acid (FA 16:0) accounting for 39.05 ± 0.66% and γ-linolenic acid (FA 18:3 *n*–6) for 31.32 ± 0.30%. Nutritional quality indices included atherogenic (0.75 ± 0.03%), thrombogenic (1.32 ± 0.04%), and hypocholesterolemic/hypercholesterolemic index (0.45 ± 0.03%). Acetonic extracts contained 78.66% chlorophyll-*a*, 16.18% total carotenes, and 5.11% xanthophylls. The aqueous extract showed the following contents ($\mu g\ mg^{-1}$): phenolics with 15.64 ± 1.76 as gallic acid equivalents, sugars with 7.19 ± 0.75 of glucose equivalents, chlorophyll-*a* with 5.84 ± 1.87 $mg\ g^{-1}$ of dry extract, and total chlorophylls with 10.02 ± 2.06 $mg\ g^{-1}$ of dry extract. Phycobiliprotein content revealed concentrations of phycocyanin, allophycocyanin, and phycoerythrin at 84.88 ± 8.29, 53.11 ± 6.21, and 20.25 ± 3.12 $\mu g\ mg^{-1}$ of dry extract, respectively. The nutritional profile of LEGE 231238 is comparable to that of other Spirulina strains currently available on the market, supporting its suitability as a food supplement. This work represents the first published report of an *L. platensis* isolate from Bolivia and underscores its strong potential for commercial exploitation.

Keywords: fatty acids, Lege-CC, nutrition, phylogeny, pigments, phenols, phycobiliproteins, protein, Spirulina, sugar

1. Introduction

Cyanobacteria, previously referred to as blue-green algae, are among the oldest oxygenic photosynthetic organisms on Earth and play a crucial role as primary producers with a global distribution [1]. Their high biotechnological potential has been widely documented, largely attributed to their rapid growth rate, ability to synthesize pigments such as chlorophyll and phycocyanin (PC), and capacity to accumulate diverse bioactive compounds under variable environmental conditions [2, 3]. These metabolites exhibit diverse applications in human health, encompassing cardioprotective, anti-inflammatory, and antioxidant effects [4–6].

In human nutrition, species of the genus *Limnospira* Nowicka-Krawczyk, Mühlsteinová and Hauer (commercially known as Spirulina) deserve special attention. They are an exceptionally rich source of protein, representing 70% of dry weight (DW), considerably higher than conventional sources such as meat, which contains around 22% [7–9]. Moreover, *Limnospira* is a source of polyunsaturated FA (PUFAs), particularly gamma-linolenic acid, along with a wide range of vitamins (A, D, E, K, C, and B-complex) and micronutrients such as calcium, iron, magnesium, potassium, and selenium. It is also rich in bioactive compounds, including PC, known for its strong antioxidant activity, and several carotenoids, including β-carotene, canthaxanthin, lutein, astaxanthin, and zeaxanthin, contributing to its nutraceutical value. Its cell wall contains polysaccharides like peptidoglycans and hemicellulose, potentially acting as prebiotics with immunomodulatory properties [10–13].

This nutritional value is historically known and can be traced back to the sixteenth century, when the Aztecs harvested *Limnospira* biomass from Lake Texcoco to produce dried cakes. Comparable practices have also been reported around Lake Chad in Africa, highlighting the long-standing nutritional and cultural significance of this cyanobacterium [2, 10, 14, 15]. In recent decades, its use in the food industry has expanded considerably, with cultivation being carried out primarily in open raceway ponds, which yield dry powder biomass, and in closed photobioreactors that enable the harvesting of fresh biomass [10, 16–18]. Open systems are economically feasible but face challenges such as temperature fluctuations, contamination, nutrient balance, and production optimization. In contrast, closed systems allow for tight control over variables such as temperature, pH, light intensity, and nutrient concentration, improving yield and biomass composition while minimizing contamination risks [13, 15, 19]. Owing to its simple cultivation and the high nutritional value of the final product, *Limnospira* has become one of the most commercialized “microalgae” worldwide [20].

As a result, *Limnospira* has been incorporated into a wide variety of food products, including energy bars, pasta, bread, cookies, beverages, nutritional supplements, infant formulas, and snacks such as yogurt, cheese, and ice cream [2, 11, 21–24]. The addition of *Limnospira* to these products not only increases their nutritional value but also improves oxidative stability, extends shelf life, and provides a characteristic green color due to the abundance of phycobiliproteins (PBPs), particularly PC [25]. Beyond food applications, *Limnospira* also shows strong potential in the development of functional foods, which are characterized by delivering health-promoting effects that go beyond basic nutrition. Its metabolites have been shown to help reduce blood lipids, combat diabetes and obesity, and possess antiviral, anti-inflammatory, immunostimulatory, and anticancer properties [2, 7, 8, 21].

Owing to this economic importance, *Limnospira* is one of the most studied cyanobacteria, and its taxonomic classification has experienced multiple revisions over time, persisting into the present. The species within this genus were initially described under *Spirulina* Gomont, such as *S. fusiformis* Voronichin [26] and *S. maxima* Geitler [27], or under *Arthrospira* Gomont, such as *A. platensis* Gomont [28]. In a major taxonomic revision of *Arthrospira*, Komárek and Lund, in 1990 [29], reassigned *S. maxima* and *S. fusiformis* to the planktonic forms of *Arthrospira* (with aerotopes), renaming them *A. maxima* Setchell et Gardner and *A. fusiformis* Komárek et Lund. Later, based on 16S rRNA gene analyses, Nowicka–Krawczyk et al., in 2019 [30], separated these planktonic (with aerotopes) forms of *Arthrospira* into the genus *Limnospira*. More recently, in 2023, Santos et al. [31] revised the taxonomy of *A. platensis*, ultimately transferring this species to *Limnospira* and establishing a reference strain for it.

Taken together, studies on *Limnospira,* its consumption, and incorporation into novel functional food formulations represent promising strategies to address global nutritional demands, promote sustainable diets, and diversify protein sources under climate change and demographic pressures. Moreover, the isolation of new *Limnospira* strains is particularly relevant for sustainable development, especially in developing countries. Accordingly, this study reports a newly isolated *Limnospira* strain from Bolivia and evaluates its nutritional value, highlighting its economic potential as a promising alternative resource.

2. Materials and methods

2.1 Cyanobacteria biomass production and morphological analysis

The strain *Limnospira platensis* LEGE 231238 was isolated in August 2023 from Alalay Lake, Bolivia (17°24′28″ S, 66°08′11″ W), a hypertrophic brackish water body [32]. Currently, the strain is maintained at the Blue Biotechnology and Ecotoxicology Culture Collection (LEGE-CC) at the Interdisciplinary Centre of Marine and Environmental Research (CIIMAR/CIMAR LA), Matosinhos, Portugal, in Z8 liquid medium supplemented with 25 g L^{-1} ($Z8_{25}$) of Synthetic Sea Salt (Tropic Marine, Berlin, Germany) and 10 μg mL^{-1} of cyanocobalamin (Kotai, 1972) under the following conditions: 12/12 h (light:dark cycle) with 25 μmol photons m^{-2} s^{-1} and 19°C.

For biomass production, a culture volume of 20 L was prepared using a 1 L inoculum and maintained under continuous aeration for 20 days, with a controlled temperature of 21 ± 2°C, a 16/8 h light/dark photoperiod, and a light intensity of 20 μmol m^{-2} s^{-1} (AsenseTek Lighting Passport, Biosystems, United Kingdom) provided by fluorescent lamps (Arcadia, United Kingdom). Collected biomass was concentrated using a 20 μM nylon mesh (Altmann Analytik, Munich, Germany), and an additional washing step with deionized water was performed to remove salt. The collected biomass was dried in a freeze dryer (Lyoquest-55, Telstar Technologies, S. L., Spain) and stored away from light at − 20°C prior to extract preparation.

For the morphological analysis, LEGE 231238 was observed, microphotographed, and analyzed using the MOTIC Panthera light microscope with a camera and dedicated software (MOTIC Europe, Barcelona, Spain). Thirty cell measurements were performed across various areas of slide preparations.

2.2 DNA extraction, PCR amplification, sequencing, and phylogenetic analysis

Cyanobacterial cells were collected from the culture, and total genomic DNA (gDNA) extraction, PCR amplification, and sequencing were performed following the study by Oliveira et al. [33]. The phylogenomic analyses were conducted according to the study by Hentschke et al. [34], aligning the 16S rRNA gene sequence of LEGE 231238 with *Limnospira* reference strains identified by Santos et al. [31], as well as with other Microcoleaceae.

2.3 Chemical analysis

2.3.1 Extracts preparation

A sequential extraction was carried out from the dry biomass using solvents of different polarities (acetone and water) according to Silva et al. [35] and De Rosa et al. [36], with some modifications. The acetone extract was prepared by mixing 0.5 g of dried biomass with 40 mL of 100% high-performance liquid chromatography (HPLC)-grade acetone. The suspension was sonicated at room temperature in an ultrasonic homogenizer (Sonoplus, probe MS 72, Bandelin, Berlin, Germany) for 5 minutes, with a pulse on for 59 seconds and a pulse off for 3 seconds (amplitude 95%), in order to help break the cell walls and release the cell content. The mixture was centrifuged (Megafuge TM 16 R, Thermo Fisher Scientific, Bremen, Germany) at 5,000 × *g* at 4°C for 5 minutes, and the extraction process was repeated three times for the same biomass. After completing four extraction sequences, the supernatants were evaporated (BUCHI R-210 Rotary Evaporator, Switzerland). The resulting pellet from the acetone extraction was left to dry overnight in the fume hood. After completely drying, the biomass was extracted with 40 mL of ultrapure water (aqueous extract), following the same procedure as before. The supernatants of the aqueous extract were combined, frozen, and lyophilized (Lyoquest-55, Telstar Technologies, S. L., Spain). The dry extracts (DEs) were kept at - 20°C until chemical analysis.

2.3.2 Carotenoid and chlorophyll profiling by HPLC with photodiode array detector

Carotenoid and chlorophyll analyses ($n = 4$) were determined from acetonic extracts (10 mg mL^{-1}) using a Waters Alliance 2695 HPLC system equipped with a photodiode array detector (PDA, Waters, USA), following a previously described protocol [37] and the methodology described by Oliveira et al. [38], with the following modifications: quantification at 450 nm using external calibration with lutein, zeaxanthin, chlorophyll-*a*, and β-carotene (Extrasynthese, Genay, France). Oxygenated derivatives of β-carotene were expressed as zeaxanthin, β-carotene derivatives as β-carotene, and chlorophyll-*a* derivatives as chlorophyll-*a*. Myxoxanthophyll was identified with an authentic standard solution (DHI, Horsholm, Denmark) and quantified as zeaxanthin. Calibration curves were constructed using four concentrations of standards covering the observed range in samples. Calibration parameters, including r^2 values, limits of detection ($LOD = 3S_0/b$), and quantification ($LOQ = 10S_0/b$, where S_0 is the noise standard deviation and b is the calibration slope), are reported in **Table 1**.

Standards	Calibration curve	r^2	LOD (μg mL^{-1})^a	LOQ (μg mL^{-1})^b
Lutein	y = 523141565217x – 1977200435	0.9930	5.657	17.142
Chlorophyll-*a*	y = 52427733333x + 26570	0.9981	8.909	26.996
Zeaxanthin	y = 662580827826x + 1302254348	0.9997	1.253	3.797
β-Carotene	y = 34727408,0000x – 96390	0.9992	3.690	11.183

[a]*LOD: Limit of detection.*
[b]*LOQ: Limit of quantification.*

Table 1.
Calibration curves used for pigment quantification.

2.3.3 Spectrophotometric analysis of aqueous extracts

The aqueous extract was dissolved in water to a final concentration of 10 mg mL^{-1}, and PBPs were quantified by measuring absorbance at 562, 615, and 652 nm using a 1cm path length cuvette. This quantification (n = 4) was performed and calculated using the following Eqs. 1, 2, and 3 [39]. Similarly, chlorophyll-*a* and total chlorophylls were quantified by measuring the absorbance at 630, 647, 664, and 691 nm (in a quartz cell with 1 cm of optical path), using acetone-water 9:1 to a final concentration of 10 mg mL^{-1}. This quantification (n = 4) was performed and calculated using the following Eqs. 4 and 5 [40]. The results were expressed in micrograms of the respective PBP and chlorophyll per milligram of DE.

$$\text{Phycocyanin (PC)} = \frac{A_{615\,nm} - 0.474\,\text{x}\,A_{652\,nm}}{5.34} \tag{1}$$

$$\text{Allophycocyanin (APC)} = \frac{A_{652\,nm} - 0.208\,\text{x}\,A_{615\,nm}}{5.09} \tag{2}$$

$$\text{Phycoerythrin (PE)} = \frac{A_{562\,nm} - 2.41\,\text{x}\,PC - 0.849\,x\,APC}{9.62} \tag{3}$$

$$\text{Chlorophyll-}a\ (\text{Chl-}a) = -\ 0.3319\ A_{630\,nm} - 1.7485\ A_{647\,nm} + 11.9442\ A_{664\,nm} - 1.4306\ A_{691\,nm} \tag{4}$$

$$\text{Total chlorophyll (Chl)} = 21.3877\ (A_{630\,nm} + 10.3739\ (A_{647\,nm}) + 5.3805\ (A_{664\,nm}) + 5.5309\ (A_{691\,nm}) \tag{5}$$

2.3.4 Total phenolic content analysis of aqueous extracts

Total phenolic content (TPC) was measured using the Folin–Ciocalteu colorimetric assay, following the method described by Barroso et al. [41], as modified by Morone et al. [42]. Previously, we prepared the gallic acid (GA) solution at 1 mg mL^{-1} (Sigma, 97.5%–102.5%) as a standard, the Na_2CO_3 (VWR, 99.5%–100.5%) solution at 75 g L^{-1}, and ethanol (for clinical diagnosis, PanReac, AppliChem, ITW Reagents, Barcelona, Spain) at 80%. Calibration curves (y = 2.7771x – 0.0164, r^2 = 0.9998) were performed using seven different concentrations of the standard solution. To prepare the blank, in Eppendorf tubes (n = 4), we added 25 μL of the standard solution (GA), 25 μL of ethanol (80%), 200 μL of Na_2CO_3 solution, and 500 μL of deionized water. For the Folin reaction, we used Folin–Ciocalteu reagent (Sigma-Aldrich, St. Louis, MO, USA) to replace ethanol (80%). For the cyanobacterial strain, the aqueous extract was dissolved in ethanol (80%) to obtain a final concentration of 5 mg $mL^{-1,}$ replacing the standard solution. The incubation and measurement procedures were carried out according to Morone et al. [37]. TPC was expressed as milligram GA equivalents (GAE) per gram of DW. Each assay was performed in four independent replicates (n = 4).

2.3.5 Total sugars quantification of aqueous extracts

The determination was performed according to Masuko et al. [43] and DuBois et al. [44], with some modifications. Previously, the phenol solution (20%) (Acros Organic, Geel, Belgium) and glucose monohydrate (1 mg mL^{-1} (Sigma Merck G8270, ≥ 99.5%) solution, used as a standard, were prepared. Calibration curves ($y = 1.4616x + 0.0252$, $r^2 = 0.9985$) were generated using six different concentrations of the standard solution. In an Eppendorf tube, we added 100 μL of the standard solution, 500 μL of H_2SO_4 (≥ 95%, Fisher Chemical), and 100 μL of the phenol solution (20%). For the blank, we used ethanol (80%), and for the samples, the aqueous extract was dissolved in ethanol (80%) to obtain a final concentration of 5 mg $mL^{-1,}$ replacing the standard solution. Finally, the microplate was prepared with 200 μL of each Eppendorf tube (blank, standard, sample blank, and samples) and incubated for 30 minutes. Then, the total sugar content was determined by measuring the absorbance at 490 nm with a spectrophotometer (Biotek, Bad Friedrichshall, Germany). The results were expressed in μg glucose equivalents *per* mg dry extract (DE). The experiment was carried out in quadruplicate (n = 4).

2.4 Biomass proximate composition and lipid extraction procedure

Procedures were conducted at Greencolab (https://www.greencolab.com/, accessed on 01 April 2025), following the methodology described in Oliveira et al. [38]. The proximate composition of cyanobacterial biomass was determined by quantifying protein, ash, and carbohydrate contents, with carbohydrates estimated by difference from total DW [45]. Lipids were extracted following the Folch method [46, 47], with minor modifications, as described in detail by Oliveira et al. [38]. Lipid content was expressed as % DW using Eq. 6, and results are reported as mean ± standard deviation (n = 5).

$$\text{Lipid content yield}\left(\%\ \text{DW}\frac{\text{W}}{\text{W}}\right)=\frac{\text{Weight of the lipid extract (g)}}{\text{Weight of biomass (g)}}\times 100 \qquad (6)$$

2.4.1 FA analysis by gas chromatography–mass spectrometry (GC–MS)

FA methyl esters (FAMEs) were prepared following the method described in literature [48] and detailed in the study by Oliveira et al. [38]. Results were expressed as mg FA g^{-1} DW, and nutritional quality indexes (AI, TI, and h/H) were calculated according to Eqs. 7–9.

$$AI = \frac{[FA\ 12{:}0 + (4 \times FA\ 14{:}0) + FA\ 16{:}0]}{[\sum MUFA^{*} + \sum(n\text{-}6) + \sum(n\text{-}3)]} \quad (7)$$

$$TI = \frac{[FA\ 14{:}0 + FA\ 16{:}0 + FA\ 18{:}0]}{[(0.5 \times FA\ 18{:}1) + (0.5 \times \sum Other\ MUFA) + (0.5 \times \sum(n\text{-}6)) + (3 \times \sum(n\text{-}3)) + (\frac{\sum(n\text{-}3)}{\sum(n\text{-}6)})]} \quad (8)$$

$$(h/H) = \frac{[FA\ 18{:}1\ (n\text{-}9) + FA\ 18{:}2\ (n\text{-}6) + FA\ 18{:}3\ (n\text{-}3) + FA\ 20{:}4\ (n\text{-}6) + FA\ 20{:}5\ (n\text{-}3)]}{[FA\ 14{:}0 + FA\ 16{:}0]} \quad (9)$$

* Monounsaturated FA

3. Results

3.1 Phylogeny and taxonomy

The ML and BI phylogenetic analyses (see **Figure 1**) placed strain LEGE 231238 within the *Limnospira* clade with strong statistical support (ML = 98, BI = 1). It clustered with the reference strain *L. platensis* CCIBt3254 [31], together with strains *L. platensis* CCIBt3335 and CCIBt3336. The 16S rRNA gene identity between LEGE 231238 and these strains ranged from 99.8% (vs. CCIBt3336) to 100% (vs. CCIBt3254 and CCIBt3335).

Morphologically, strain LEGE 231238 exhibits solitary trichomes with regular coils, not constricted at the cross walls, and lacking a mucilaginous envelope. The spacing between spirals ranges from 30 to 51 μm, while spiral height varies from 25 to 31 μm. Trichome width ranges from 8 to 10 μm, and the cells are shorter than wide, measuring 4–7 μm in length. The apical cells are rounded, often with calyptra. The cell content is blue-green, with numerous aerotopes (see **Figure 2**). The planktonic habit of LEGE 231238, originating from a brackish water body, is consistent with the species' ecology [31].

Based on that, it was evident that the molecular, morphological, and habitat analyses are in agreement and confirm the identification of LEGE 231238 as *Limnospira platensis*.

3.2 Chemical composition analysis

The chemical composition of the biomass, acetonic, and aqueous extract from *L. platensis* LEGE 231238 was characterized. Additionally, data from previous studies with Spirulina were included for comparison (see **Tables 2** and **3**). Analysis of the acetonic extracts using HPLC-PDA revealed the presence of several chlorophyll and carotenoid compounds, yielding a good chromatographic resolution, as shown in **Figure 3**. The most abundant pigment was chlorophyll-*a* (283.53 ± 11.01 mg g^{-1} DE) and chlorophyll-*a* derivative (52.71 ± 3.95 mg g^{-1} DE). Among the carotenoids, the most abundant were β-carotene (45.63 ± 0.98 mg g^{-1} DE) and zeaxanthin (16.56 ± 1.70 mg g^{-1} DE). Other carotenoids detected included myxoxanthophyll (5.43 ± 0.50 mg g^{-1} DE), an unidentified xanthophyll (4.28 ± 0.24 mg g^{-1} DE), β-cryptoxanthin (6.79 ± 0.60 mg g^{-1} DE), an unidentified carotenoid (3.30 ± 0.01 mg g^{-1} DE), and a β-carotene isomer (9.13 ± 0.80 mg g^{-1} DE). When considering the total pigment composition, chlorophylls accounted for 78.66% of the total identified pigments, while carotenes represented 16.18%, and xanthophylls 5.11% (see **Figure 4**).

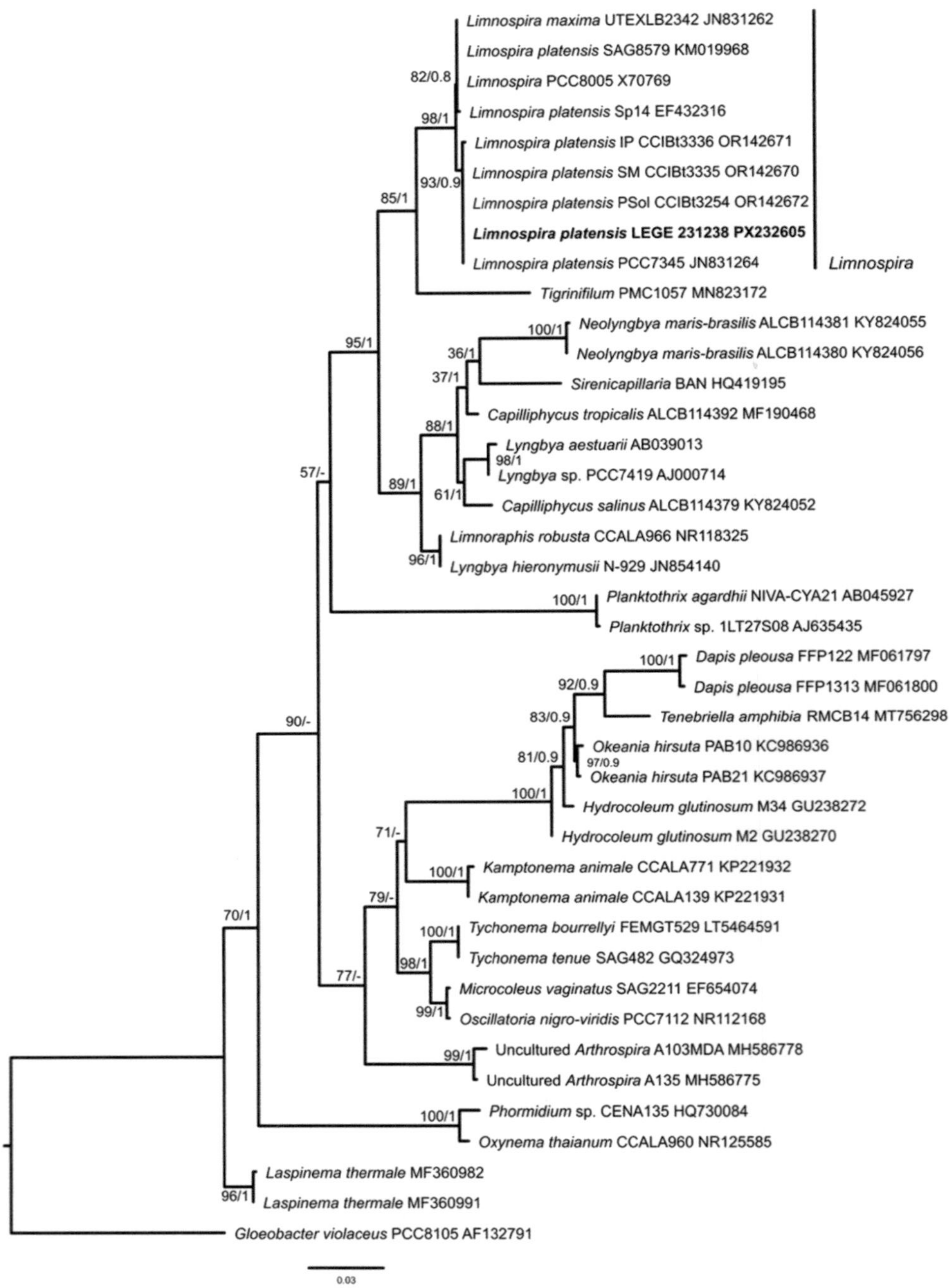

Figure 1.
Phylogenetic tree with 41 sequences, inferred from the 16S rRNA gene using maximum-likelihood analysis. Bootstrap values and posterior probabilities are displayed at the nodes. The strain L. platensis *LEGE 231238 is in bold.*

The aqueous extract was characterized by spectrophotometry for its total phenolic content, total sugars, chlorophylls, and PBPs. The phenolic content was 15.64 ± 1.76 μg of GAE mg^{-1} of DE, and the total sugar content was determined to be

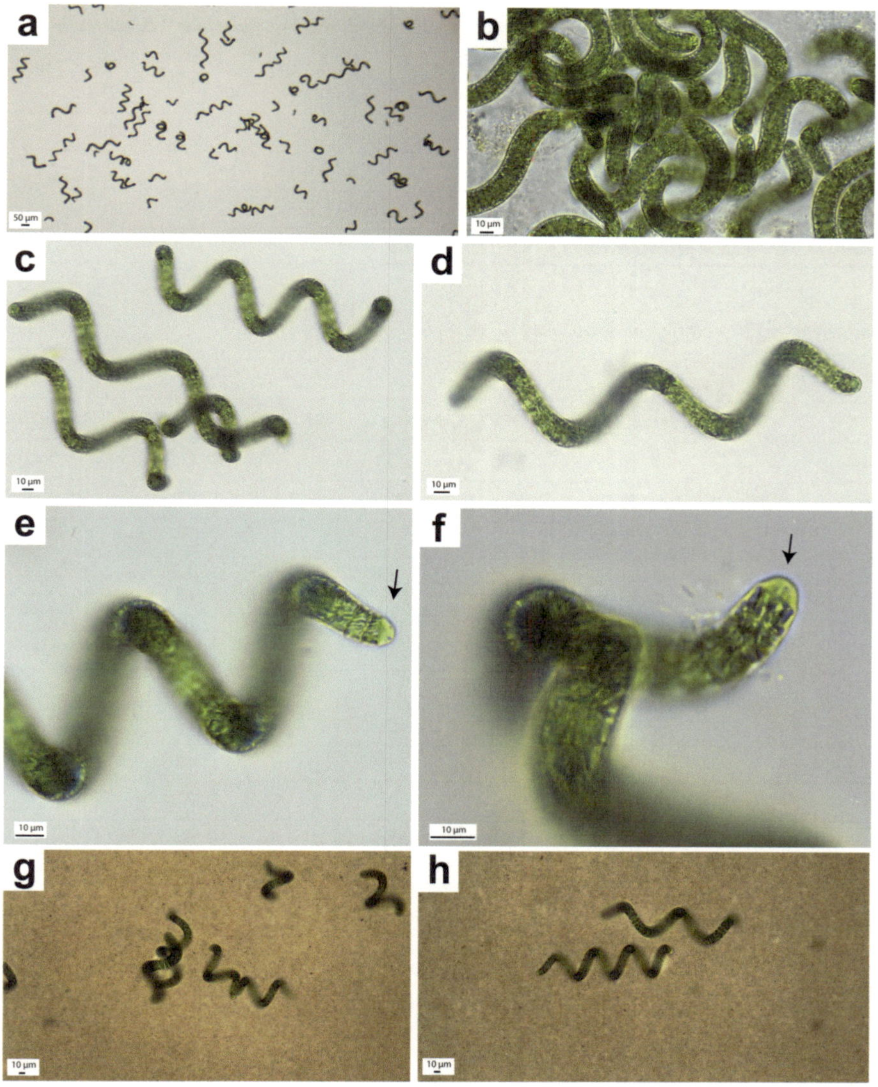

Figure 2.
Microphotographs of L. platensis *LEGE 231238. (a) General overview of trichome arrangement; (b) general overview showing trichomes with aerotopes; (c, d) details of trichome coiling; (e, f) trichomes with calyptra, pointed out by the black arrows and aerotopes; (g, h) Slides prepared with China ink, showing the lack of mucilaginous envelopes.*

7.19 ± 0.75 μg of glucose equivalents mg^{-1} of DE (based on the calibration curve obtained in the assay). Regarding the chlorophyll-*a* and total chlorophylls content, the values were 5.84 ± 1.87 (3.4%) and 10.02 ± 2.06 (5.8%) μg mg^{-1} of DE, respectively. The PBPs profile revealed concentrations of 84.88 ± 8.29 μg mg^{-1} of DE for PC (48.8%), 53.11 ± 6.21 μg mg^{-1} of DE for APC (30.5%), and 20.25 ± 3.12 μg mg^{-1} of DE for PE (11.6%).

The proximate composition of cyanobacterial biomass revealed a high protein content, with 45.40 ± 0.54% of DW. Carbohydrates were the second most abundant component, with 27.84 ± 0.25% of DW, followed by ash content with 19.90 ± 0.32%

	L. platensis LEGE 231238	*Spirulina* sp. [2]	*S. platensis* [9]	*A. platensis* [49]	*Spirulina* sp. [50]	*S. platensis* [48]	*Spirulina* [51]
Total protein	45.4	57.5	46–63	23.8–62.9	48.2	67.8	62.8
Total lipids	6.9	7.7	4–9	1.5–8.1	5.3	1.6	6.9
Carbohydrates	27.8	23.9	8–14	3.6–15.8	14.2	7.1	–
Ash	19.9	6.2	–	8.2–18.11	10.1–14.6	11.7	7.5

The values are expressed in pecentage (%).

Table 2.
Macronutrient profile of different Spirulina strains.

	L. platensis LEGE 231238	*Spirulina* sp. [47]	*S. platensis* [52]	*A. platensis* [49]
FA 14:0	1.23	0.1	0.28	0.42–3.65
FA 16:0	39.05	38.6	42.85	43.60–71.15
FA 16:1	1.45	–	–	–
FA 16:1 *n*–7	6.82	5.1	3.16	4.36–7.60
FA 18:0	0.96	5.4	1.00	0.69–3.33
FA 18:1 *n*–9	3.10	2.6	3.53	4.02–5.99
FA 18:1	1.16	–	–	–
FA 18:2 cis (*n*–6)	14.92	21.4	14.43	7.86–20.15
FA 18:3 *n*–6	31.32	23.3	22.45	2.73–4.50
AI	0.75	0.7	0.89	–
TI	1.32	1.6	0.28	–
PUFA	46.23	44.8	40.6	10.59–24.52
SFA	41.24	44.2	50.1	62.94–80.31
MUFA	12.53	10.9	8.5	9.11–12.58
n–3 PUFA	–	–	2.9	0.01–0.20
n–6 PUFA	46.23	44.6	37.7	10.59–24.48
h/H	0.45	0.6	1.0	–

Values are expressed in relative abundance (%) and represent the mean of five analytical samples.

Table 3.
Fatty acid profile identified in the total lipid extract of the cyanobacteria strain L. platensis *LEGE 231238.*

of DW and lipid content with 6.86 ± 0.18% of DW. Elemental CHN analysis showed that carbon (C) represented 38.83 ± 0.28% of DW, hydrogen (H) 6.47 ± 0.08% of DW, and nitrogen (N) 7.26 ± 0.09% of DW, and the C:N ratio was 5.54 ± 0.17% of DW (see **Figure 4**). These values are in agreement with data from previous studies on Spirulina strains, as shown in **Table 2**.

The FA composition of the lipid extract was determined to assess its lipid profile and potential nutritional and biotechnological applications. This analysis allowed the identification of nine FAs across *L. platensis* LEGE 231238 (see **Figure 5**) and revealed a predominance of saturated FA (SFA) with 41.24 ± 0.72% and

DOI: http://dx.doi.org/10.5772/intechopen.1013578

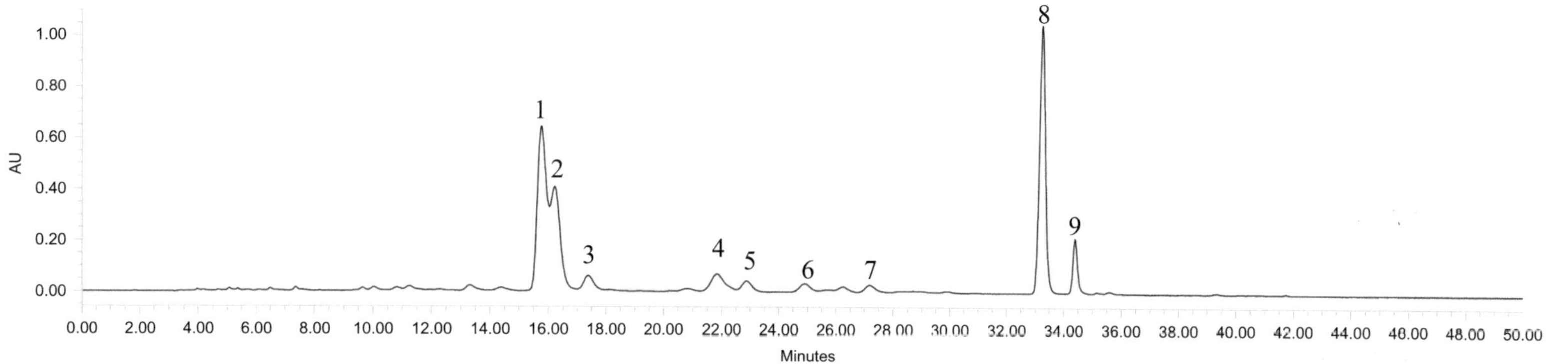

Figure 3.
Carotenoid and chlorophyll profiles of the acetonic extract of L. platensis *LEGE 231238 obtained by high-performance liquid chromatography with photo-diode array detection (HPLC-PDA), recorded at 450 nm. Chlorophyll-*a *(1), zeaxanthin (2), chlorophyll-*a *derivative (3), myxoxanthophyll (4), unidentified xanthophyll (5), β-cryptoxanthin (6), unidentified carotenoid (7), β-carotene (8), and β-carotene isomer (9).*

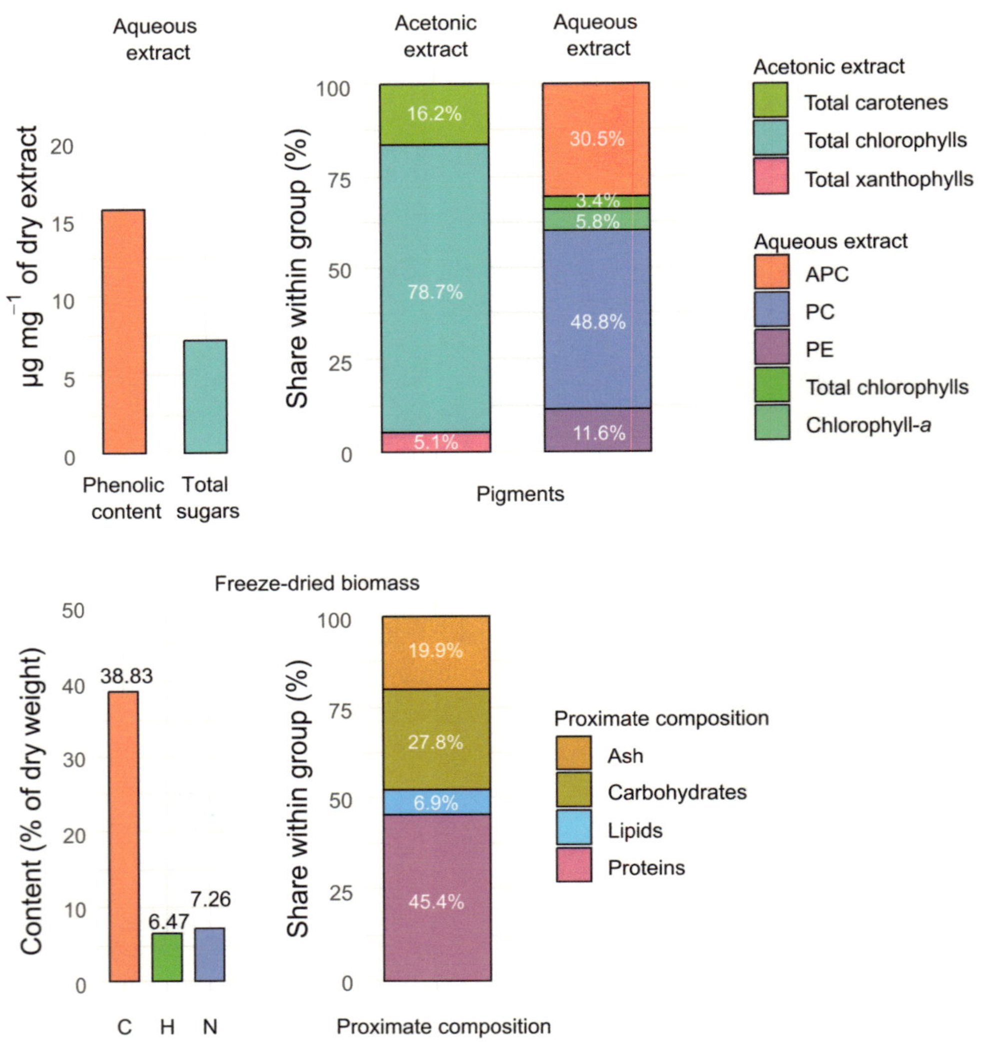

Figure 4.
Graphical representation of the composition and relative distribution of the analyzed groups. Bar charts on the left show absolute abundance, while stacked bar charts on the right display the percentage contribution of each group. Colors represent different categories.

polyunsaturated FA (PUFA) with 46.23 ± 0.60%. The most abundant FA was palmitic acid (FA 16:0), representing 39.05 ± 0.66%, followed by γ-linolenic acid (FA 18:3 *n*–6) with 31.32 ± 0.30% and linoleic acid (FA 18:2 cis *n*–6) with 14.92 ± 0.73%. Among the MUFAs (12.53 ± 0.50%), palmitoleic acid (FA 16:1 *n*–7) was the most abundant (6.82 ± 0.12%), followed by FA 16:1 and 18:1 (1.45 ± 0.12% and 1.16 ± 0.13%, respectively). Minor constituents were also identified, such as oleic acid (FA 18:1 *n*–9) with 3.10 ± 0.27%, myristic acid (FA 14:0) with 1.23 ± 0.08%, and stearic acid (FA 18:0) with 0.96 ± 0.08%. The potential health benefits of the cyanobacterial strain were also determined by AI (0.75 ± 0.03%), TI (1.32 ± 0.04%), and h/H index (0.45 ± 0.03%).

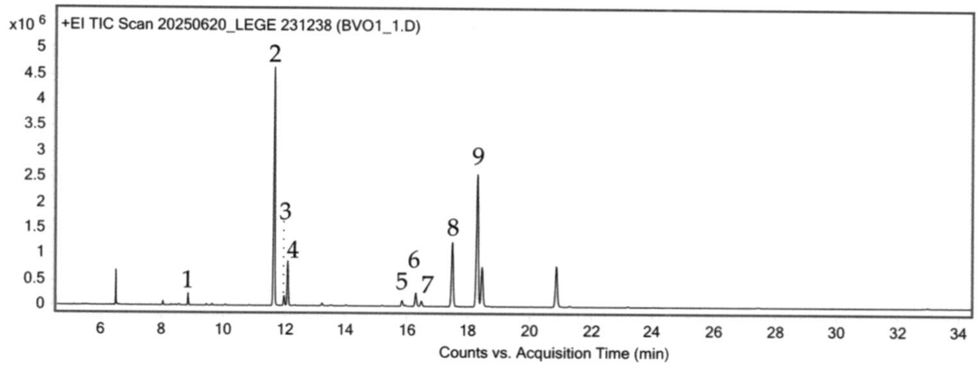

Figure 5.
FAs profile of the lipid extract of L. Platensis *LEGE 231238 obtained by gas chromatography–mass spectrometry (GC–MS). FA 14:0 (1), FA 16:0 (2), FA 16:1 (3), FA 16:1* n–7 *(4), FA 18:0 (5), FA 18:1* n–9 *(6), FA 18:1 (7), FA 18:2 cis* n–6 *(8), and FA 18:3* n–6 *(9).*

4. Discussion

The growing market interest in sustainable and alternative food sources, nutritional supplements, and bioactive compounds has positioned cyanobacteria at the forefront of research. Spirulina is renowned for its rich and diverse biochemical composition, particularly protein, carbohydrates, pigments, and essential lipids, such as *n*–6 FA, which underscore its importance as a nutritious and sustainable dietary supplement [53].

To understand the nutritional potential of *L. platensis* LEGE 231238, it is essential to evaluate the biomass composition, thereby providing an overview of its biochemical profile. Carbon and nitrogen metabolism are essential for biological systems, as they are involved in the synthesis of energy carriers, proteins, and pigments, and are tightly interconnected in various organisms. In plants and prokaryotes, this coupling is primarily driven by two main factors [54–57]. First, carbon (C) and nitrogen (N) are among the most abundant elements in cells, requiring coordination to avoid metabolic inefficiencies. Secondly, nitrogen assimilation depends on the availability of C skeletons for biosynthesis; therefore, a deficiency or excess of one element significantly influences the metabolism of the other [57]. Such regulation can occur at multiple levels, from allosteric control of nutrient assimilation proteins to mechanisms regulating the expression of their genes [58–60]. Hydrogen content (H) is typically used to determine the calorific value, which reflects its energy potential [61].

These CHN values observed in *L. platensis* LEGE 231238 are consistent with the typical ranges reported for cyanobacterial biomass and reflect a high nutritional value [62]. The elevated N content supports the high protein level observed and reinforces its potential as a sustainable source of bioactive and nutritional compounds. Proteins are the primary nitrogen-containing biomolecules in microalgae and cyanobacteria. Furthermore, the C:N ratio was approximately 5.35 ± 0.2%, which is indicative of protein-rich biomass [63]. A low C:N ratio is generally characteristic of nitrogen-rich cells and is considered desirable for nutritional and biotechnological applications, particularly in the food and feed industries [63].

Regarding that, Spirulina is globally recognized for its high protein content and nutritional value. When comparing the protein content obtained herein (45.40 ± 0.54%) to other cyanobacterial genera, such as *Spirulina* [2], *Anabaena,* and *Aphanizomenon* [9], with values of 57.5%, 43%–56%, and 62%, respectively, we notice that, even being in a similar range, the protein content of *L. platensis* LEGE 231238 was slightly lower. However, the protein content of this strain is within the Spirulina range. Aouir et al. [49] demonstrated that *L. platensis* (identified as *Arthrospira platensis*; in powder and lyophilized form) from different countries and cultivated in different media has a protein content ranging from 23.78% to 62.86%. Another study conducted by Rodriguez de Marco et al. [48] identified the protein from *L. platensis* (as *S. platensis*) from Cuba with 67.82 ± 1.71%. In general, it seems that the protein content is correlated to cultivation conditions, as well as different culture media [49].

Another important parameter is the lipid content, as it is key to evaluating the nutritional potential of cyanobacterial biomass. Spirulina has been widely recognized for its potential health benefits, particularly in the context of obesity and metabolic disorders [64]. Preclinical and clinical studies have demonstrated its efficacy in improving various health parameters, such as reducing body weight, blood pressure, and plasma triglycerides, while also exhibiting antioxidant and cardiomyocyte-protective activity [4, 65, 66]. From a nutritional perspective, Spirulina is particularly valued for its FA profile, especially as a rich source of *n*–6 PUFAs. These FAs cannot be synthesized by the human body and must be obtained through the diet [67]. While most modern diets are deficient in *n*–6 PUFAs, the inclusion of supplements like Spirulina can help rebalance lipid intake [68]. Spirulina has been identified as one of the best whole food sources of *n*–6, offering an accessible option for dietary supplementation [69, 70]. Given the alarming global rise in obesity-related diseases, with millions of deaths and significant burdens on public health systems, the incorporation of Spirulina into daily nutrition may offer a promising and natural strategy to support weight management and improve metabolic health [4, 71, 72].

In our study, *L. platensis* LEGE 231238 showed similar lipid content (6.86%) to the commercial Spirulina, as described in **Table 2** (1.5%–8.1%). Differences in lipid content among the cyanobacteria are influenced by diverse factors, as reported in the study with *Synechocystis* sp. PCC 6803 [73], also including nutrient availability, light intensity, and oxidative stress, among others. It is important to note that total lipid levels can vary between species [9] and are also affected by the extraction method applied [74]. Conde et al. [47] identified in *Limnospira* sp. (as *Spirulina* sp.) a higher lipid content (10.7% of biomass) when compared to our cyanobacterial strain. The total PUFA content in our strain (46.23%) also agrees with the typical ranges reported for *Limnospira* (as *Spirulina*), in a study conducted by Conde et al. and Selina et al. [47, 52]. Another study by Aouir et al. [49] identified lower PUFA content (10.59%–24.52%) in *L. platensis* (as *A. platensis*). Notably, in some studies, the PUFA profile was entirely dominated by *n*–6 FA, with no detectable *n*–3 PUFAs [47, 49].

Nutritional indices such as the AI and TI are valuable parameters for assessing the potential health effects of dietary lipids [75]. In our study, the values of AI (0.75%) and TI (1.32%) are similar (see **Table 3**) to those reported in commercial Spirulina [47, 52]. Although the TI was moderately high, this is expected due to the elevated SFA content, particularly palmitic acid, which is the most abundant in our cyanobacterial strain. The *n*–6/*n*–3 PUFA ratio was not possible to calculate, given the absence of detectable *n*–3 PUFAs. The PUFA/SFA index is used to evaluate the nutritional value of foods and the impact of diet on cardiovascular disease, with PUFAs having a beneficial effect on health, whereas SFAs have a positive impact on high cholesterol levels [76, 77]. *L. platensis* LEGE 231238 demonstrated a similar PUFA/SFA index value (1.12%) to those reported for *Sargassum tenerrimum* (1.18%) [78], *Concholepas concholepas* (1.16%) [79], *Scomberomorus cavalla* (1.18%) [80], and chicken (0.308–2.042%) [81]. The PUFA/SFA index of *L. platensis* LEGE 231238 indicates a favorable FA profile, since high PUFA/SFA ratios are generally associated with improved nutritional quality and potential health benefits, particularly in reducing the risk of cardiovascular diseases [82, 83].

In another biotechnological context, Spirulina is renowned for producing a broad spectrum of bioactive compounds, such as chlorophylls, carotenoids, PBPs, phenolic compounds, and carbohydrates, each with valuable nutritional and biotechnological applications. Actually, there is a high demand for antioxidant compounds as food additives to help protect against oxidative stress and reduce the risk of developing non-communicable diseases (NCDs) [84].

L. platensis LEGE 231238 exhibited a remarkable pigment profile, with high levels of chlorophyll-*a* (283.53 ± 11.01 mg g^{-1} DE) and β-carotene, surpassing values previously reported for cyanobacteria [85]. Chlorophyll-*a*, the dominant photosynthetic pigment, plays a pivotal role not only in photosynthesis but also in human health, demonstrating antioxidant, anti-inflammatory, lipid-reducing, and even potential anti-obesogenic and antimutagenic activities [86, 87]. Carotenoid dietary intake and their accumulation in tissues have been positively correlated with a lower risk of chronic diseases, for which the recognized antioxidant activity of these compounds strongly contributes. β-carotene is the predominant carotenoid in the human diet, serum, and human body tissues, serving as the main source of vitamin A for humans. This carotene has a disease-protective role, being inversely correlated with the risk of ophthalmic, metabolic, and cardiovascular diseases and certain cancers [88–90]. On the other hand, zeaxanthin, together with *meso*-zeaxanthin and lutein, is a macular carotenoid, recognized for its importance in the protection of the retina, possibly preventing and/or slowing the progression of blindness [91].

Nevertheless, it is not always possible for the general population, especially the elderly, to consume adequate amounts of these pigments daily through fruits and vegetables. For this reason, the inclusion of superfoods like Spirulina in the human diet can be advantageous in terms of the stability and bioavailability of bioactive compounds of interest for human health.

PBPs are water-soluble pigment proteins that play a crucial role in harvesting light for photosynthesis in cyanobacteria, and the major types found in Spirulina include PC, APC, and PE [92, 93]. PC is the most abundant and extensively studied due to its distinctive blue color and growing interest in the food, cosmetic, pharmaceutical, and diagnostic industries [49, 94, 95]. C–PC, a PBP derived from *Aphanizomenon flos-aquae,* has been utilized in the formulation of NaturCyanin Bioactive® (CAS Noº. 11016–15-2) as a natural blue colorant alternative to synthetic dyes. PC has also demonstrated significant antioxidant properties and protective effects against oxidative stress-induced cellular damage [96]. APC is commonly used as a fluorescent marker due to its strong far-red emission and quantum efficiency [97]. PE, similarly to PC, is used as a pink colorant and has applications in fluorescence microscopy and imaging [98]. Our findings show that the values of PC, PE, and APC (84.88%, 20.25%, and 6.21%, respectively) are higher than those reported by [49], where *A. platensis* was identified with PC (1.80%–8.70%), APC (1.10%–7.20%), and PE (1.30%–3.3%). Kumar (2011) [78] also identified PC (5.38%–7.73%), APC (2.58%–3.46%), and PE (0.63%–1.80%). When *L. platensis* LEGE 231238 is compared with another cyanobacterial genus, such as *Pseudanabaena tenuis*, the values are different for PC (2%), APC (9%), and PE (89%) [96]. Thus, *L. platensis* LEGE 231238 is worth further study regarding its biological activities and safety profile, highlighting its application in the food, feed, and pharmaceutical industries for the sustainable production of bioactive compounds.

In addition to the antioxidant and protective effects of PBPs, Spirulina also contains other classes of bioactive compounds that contribute to its functional potential. Among them, phenolic compounds have garnered increasing attention due to their potent antioxidant activity [42, 99]. These metabolites act by scavenging reactive oxygen species and preventing oxidative damage to biomolecules, thereby supporting cellular defense systems [100]. *L. platensis* LEGE 231238 showed a phenolic content of 15.64 ± 1.76 μg of GAE mg^{-1} DE. These values are lower than those reported by Aouir et al. [49] for *A. platensis*, which ranged from 19.61 to 67.52 mg GAE g^{-1} of DW. This difference may be attributed to variations in the methodologies used for the extraction of phenolic compounds. However, when the same extraction and quantification methodology was applied to different genera of cyanobacteria (1.07 to 2.45 μg of GAE mg^{-1} of DE) [42], the phenolic content observed in *L. platensis* LEGE 231238 was higher. El-Baky [101] demonstrated that the phenolic content in *S. maxima* (12.94 ± 0.93 mg GAE g^{-1}) significantly correlates with its radical scavenging capacity, reinforcing the importance of these compounds in the overall bioactivity of cyanobacterial products.

Other water-soluble metabolites are present in the aqueous extract, such as total sugars, which also contribute to the nutritional and functional value of *L. platensis* LEGE 231238. Carbohydrates play a vital role in human physiology, serving as a primary energy source, regulating blood glucose and insulin metabolism, and contributing to cholesterol and triglyceride metabolism [102]. The determination of total sugar content provides a comprehensive assessment of the glycidic profile of the aqueous extract, which is crucial for the development of dietary supplements and functional products derived from cyanobacteria. The total sugar content observed in *L. platensis* LEGE 231238 was 7.19 ± 0.75 μg of glucose equivalents per mg of DE.

In conclusion, this study confirmed the classification of strain LEGE 231238 as *L. platensis* based on both molecular and morphological analyses. The nutritional profile of this strain is comparable to that of other *Spirulina* strains currently available on the market, supporting its suitability as a food supplement. Notably, this work represents the first published report of an *L. platensis* isolate from Bolivia and underscores its strong potential for commercial exploitation. Future studies on this strain should focus on optimizing cultivation conditions and enhancing the production of its bioactive compounds.

Acknowledgements

This work was funded by the Portuguese Foundation for Science and Technology (FCT) Projects UIDB/04423/2020 and UIDP/04423/2020, and by the

WP9 - Portuguese Blue Biobank under the Blue Economy Pact - Project No. C644915664-00000026, co-funded by PRR, the Portuguese Republic, and the European Union. Further support was provided by the University of Aveiro, Fundação para a Ciência e Tecnologia (FCT), and Ministério da Ciência, Tecnologia e Ensino Superior (MCTES) through CESAM (UIDB/50017/2020 + UIDP/50017/2020 + LA/P/0094/2020), with co-funding from ERDF under Portugal 2020 and Compete 2020. Marisa Pinho was funded by national funds through the FCT - Foundation for Science and Technology, I.P., under the PhD grant via CESAM (UI/BD/153346/2022). Flavio Oliveira received PhD grants from FCT (UI/BD/06241/2021). Graciliana Lopes acknowledges support from the FCT Scientific Employment Stimulus – Individual Call (10.54499/2021.01768.CEECIND/CP1665/CT0007). The Center of Biotechnology is supported by the Swedish International Development Cooperation Agency (SIDA) *through the Horizontal Program*.

Conflict of Interest

The authors declare that there are no financial or non-financial competing interests that could be perceived as influencing the work reported in this chapter.

Consent to publication

All authors have explicitly agreed to the publication of this chapter.

Author details

Guilherme Scotta Hentschke[1†], Flavio Oliveira[1,2†], Graciliana Lopes[1], Tiago Conde[3], Tánia Melo[3], Marisa Pinho[3], Maria do Rosário Domingues[3], Hugo Pereira[4], Sara Ferreira[4], Claudia Hoepfner[5,6], Vitor Vasconcelos[1,2*] and Daniel Guzmán[5]

1 CIIMAR/CIMAR LA - Interdisciplinary Centre of Marine and Environmental Research, University of Porto, Matosinhos, Portugal

2 Department of Biology, Faculty of Sciences, University of Porto, Porto, Portugal

3 Centre for Environmental and Marine Studies, CESAM, Department of Chemistry, Santiago University Campus, University of Aveiro, Aveiro, Portugal.

4 GreenCoLab – Associação Oceano Verde, Universidade do Algarve, Faro, Portugal

5 Center of Biotechnology, Faculty of Sciences and Technology, Center of Biotechnology, Universidad Mayor de San Simón, Cochabamba, Bolivia

6 GEMA Center for Genomics, Ecology, and Environment, Universidad Mayor, Santiago, Chile

† Contributed equally to work

*Address all correspondence to: vmvascon@fc.up.pt

References

[1] Whitton BA. Ecology of Cyanobacteria II: Their diversity in space and time. Durham: Springer Science & Business Media; 2012

[2] Lafarga T, Fernández-Sevilla JM, González-López C, G A-F-F. Spirulina for the food and functional food industries. Food Research International. 2020;**137**:109356

[3] Castro EM, Shannon E, Abu-Ghannam N. Effect of fermentation on enhancing the nutraceutical properties of Arthrospira platensis (Spirulina). Fermentation. 2019;**5**(1):28. DOI: 10.3390/fermentation5010028

[4] GS H, VS H, Miguel FM. Cyanobacteria for Cardiomyocyte Protection against miocardial ischemia injury: A systematic review of animal and in vitro studies. Sci Med (Porto Alegre). 2024;**34**:1–13

[5] Favas R, Morone J, Martins R, Vasconcelos V, Lopes G. Cyanobacteria secondary metabolites as biotechnological ingredients in natural anti-aging cosmetics: Potential to overcome Hyperpigmentation, loss of skin density and UV radiation-deleterious effects. Marine Drugs. 2022;**20**:183

[6] Rodrigues L, Morone J, Hentschke GS, Vasconcelos V, Lopes G. Anti-Inflammatory activity of Cyanobacteria pigment extracts: Physiological free radical scavenging and modulation of iNOS and LOX activity. Marine Drugs. 2024;**22**:131. DOI: 10.3390/md22030131

[7] AlFadhly NKZ, Alhelfi N, Altemimi AB, Verma DK, Cacciola F, Narayanankutty A. Trends and technological advancements in the possible food applications of Spirulina and their health benefits: A review. Molecules. 2022;**27**:5584

[8] P. SM, Mendes AR, Prates JAM. Chemical composition, bioactivities, and applications of spirulina (Limnospira platensis) in food, feed, and medicine. Foods. 2024;**13**:3656

[9] Becker EW. Micro-algae as a source of protein. Biotechnology Advances. 2007;**25**:207–210

[10] Luo G, Liu H, Yang S, Sun Z, Sun L, Wang L. Manufacturing processes, additional nutritional value and versatile food applications of fresh microalgae Spirulina. Front. Nutr. 2024;**11**:1455553. DOI: 10.3389/fnut.2024.1455553

[11] Gouveia L, Batista AP, Sousa I, Raymundo A, Bandarra NM. Microalgae in novel products. In: Papadopoulos KN, editor. Food Chemistry Research developments. Hauppauge, NY, USA: Nova Science Publishers, Inc.; 2008. p. 75–112

[12] Karkos PD, Leong SC, Karkos CD, Sivaji N, Assimakopoulos DA. Spirulina in clinical practice: Evidence-Based human Applications. Evidence-Based Complementary and Alternative Medicine. 2011;**2011**:531053

[13] Marjanović B, Benković M, Jurina T, S. CT, Valinger D, G. KJ, et al. Bioactive compounds from Spirulina spp.-Nutritional value, extraction, and application in food industry. Separations. 2024;**11**:257

[14] Sinetova M, Kupriyanova E, Los DA. Spirulina/Arthrospira/ Limnospira-three names of the single organism. Food. 2024;**13**:2762

[15] Habib MAB, Parvin M, Huntington TC, Hasan MR. A review on culture, production and use of Spirulina as food for humans and feeds for domestic animals and fish. FAO Fisheries and Aquaculture. 2008;1034

[16] Gallego I, Medic N, S. PJ, K. RP, Robbens J, Vereecke E, et al. The microalgal sector in Europe: Towards a sustainable bioeconomy. New Biotechnology. 2025;**86**:1–13

[17] Spolaore P, Joannis-Cassan C, Duran E, Isambert A. Commercial applications of microalgae. Journal of Bioscience and Bioengineering. 2006;**101**:87–96

[18] Hu Q. Industrial production of microalgal cell-mass and secondary products-major industrial species. In: Richmond A, editor. Handbook of Microalgal Culture: Biotechnology and Applied Phycology. Oxford, UK:Blackwell Publishing Ltd; 2004. p. 264–272

[19] Olguín EJ. Dual purpose microalgae–bacteria-based systems that treat wastewater and produce biodiesel and chemical products within a Biorefinery. Biotechnology Advances. 2012;**30**:1031–1046

[20] Belay A. The potential application of Spirulina (Arthrospira) as a nutritional and therapeutic supplement in health management. Journal of the American Nutraceutical Association. 2002;**5**:27–48

[21] Anvar AA, Nowruzi B. Bioactive properties of Spirulina: A review. Microbial Bioactives. 2021;**4**:134–142

[22] Çelekli A, Özbal B, Bozkurt H. Challenges in functional food products with the incorporation of some microalgaea. Foods. 2024;**13**:725

[23] Safi C, Zebib B, Merah O, Pontalier PY, Vaca-Garcia C. Morphology, composition, production, processing and applications of Chlorella vulgaris: A review. Renewable and Sustainable Energy Reviews. 2014;**35**:265–278

[24] Batista AP, Niccolai A, Fradinho P, Fragoso S, Bursic I, Rodolfi L, et al. Microalgae biomass as an alternative ingredient in cookies: Sensory, physical and chemical properties, antioxidant activity and in vitro digestibility. Algal Research. 2017;**26**:161–171

[25] Villaró S, Sánchez-Zurano A, Ciardi M, Alarcón FJ, Clagnan E, Adani F, et al. Production of microalgae using pilot-scale thin-layer cascade photobioreactors: Effect of water type on biomass composition. Biomass and Bioenergy. 2022;**163**:106534

[26] Voronichin NN. K biologii mineralizovanych vodojemov Kulundinskoj stepi. Trudy Soveta Izucheniu Prirodnych Resursov, Ser Sibirsk. 1934;**8**:177–183

[27] Geitler L. Cyanophyceae. In: Rabenhorst L, editor. Kryptogamen Flora von Deutschl And, Österreichund der Schweiz. Vol. **14**. Leipzig: Akademische Verlagsgesellschaft; 1932. p. 673–1196

[28] Gomont M. Monographie des Oscillariées (Nostocacées Homocystées). Deuxième partie. - Lyngbyées. Annales Des Sciences Naturelles, Botanique. 1892;**7**:91–264

[29] Komárek J, Lund JWG. What is "Spirulina platensis. Fact? Arch Hydrobiol Suppl Algol Stud. 1990;**58**:1–13

[30] Nowicka-Krawczyk P, Mühlsteinová R, Hauer T. Detailed characterization of the Arthrospira type species separating commercially grown taxa into the new genus Limnospira (Cyanobacteria). Scientific Reports. 2019;**9**:9. DOI: 10.1038/s41598-018-36831-0

[31] Santos KRS, Hentschke GS, Ferrari G, Andreote APD, de Fiore M, Vasconcelos V, et al. Molecular, morphological and ecological studies of Limnospira platensis (Cyanobacteria), from saline and alkaline lakes, Pantanal Biome, Brazil. Frontiers in Environmental Science. 2023;**11**:1204787

[32] Hentschke GS, Semedo M, Ciancas J, Hoepfner C, Guzmán D, Rivera DS, et al. Cyanobacterial mats and their associated microbiomes in saline and freshwater lakes from the Bolivian Altiplano. Frontiers in Microbiology. 2025;**16**:1650455

[33] Oliveira F, Hentschke GS, Morais J, Silva R, Cruz P, Vasconcelos VM. Exploring the cyanobacterial diversity in Portugal: Description of four new genera from LEGE-CC using the polyphasic approach . Journal of Phycology. 2024;**60**:1285–1304

[34] Hentschke GS, Ciancas JC, Hoepfner C, Mesquita MJ, Vasconcelos V. The extremophile Eurychoronema bolivianum gen. et sp. nov. (Nodosilineales, Cyanobacteria) and Leptolyngbya aquatica comb. nov. Phycologia. 2024;**63**:510–519

[35] Silva R, Gonçalves T, Morone J, Moreira GA, Morais J, Hentschke GS, et al. Pigments profile and antioxidant potential of extremophile cyanobacteria isolated from the Mexican Volcanic Lake Chichonal. Algal Research. 2024;**81**:103578

[36] De Rosa T, Hentschke GS, Lopes G, Oliveira F, Esposito G, Teta R, et al. Tusconia apicata gen. et sp. nov.: An innovative biomass source with antioxidant and anti-inflammatory potential from thermal Spring Bagni San Filippo. Algal Research. 2025;**91**:104267

[37] Morone J, Lopes G, Morais J, Neves J, Vasconcelos V, Martins R. Cosmetic application of Cyanobacteria extracts with a sustainable vision to skincare: Role in the antioxidant and antiaging process. Marine Drugs. 2022;**20**:761

[38] Oliveira F, Conde T, Pinho M, Melo T, Hentschke GS, Urbatzka R, et al. Unveiling the biotechnological potential of Cyanobacteria from the Portuguese LEGE-CC collection through lipidomics and antioxidant and Lipid-Lowering properties. Molecules. 2025;**30**:2504

[39] Bennet A, Bogorad K. Complementary chromatic adaptation in a filamentous blue-green alga. Journal of Cell Biology. 1973;**58**:410–435

[40] Ritchie RJ. Universal chlorophyll equations for estimating chlorophylls a, b, c, and d and total chlorophylls in natural assemblages of photosynthetic organisms using acetone, methanol, or ethanol solvents. Photosynthetica. 2008;**46**:115–126

[41] Barroso MF, Ramalhosa MJ, Alves RC, Dias A, Soares CMD, Oliva-Teles MT, et al. Total antioxidant capacity of plant infusions: Assessment using electrochemical DNA-based

biosensor and spectrophotometric methods. Food Control. 2016;**68**: 153–161

[42] Morone J, Lopes G, Preto M, Vasconcelos V, Martins R. Exploitation of Filamentous and Picoplanktonic Cyanobacteria for cosmetic applications: Potential to improve skin structure and preserve dermal matrix components. Marine Drugs. 2020;**18**:486

[43] Masuko T, Minami A, Iwasaki N, Majima T, Nishimura S, Lee YC. Carbohydrate analysis by a phenol–sulfuric acid method in microplate format. Analytical Biochemistry. 2005;**339**:69–72

[44] Dubois M, Gilles K, Hamilton J, Rerbers PA, Smith F. A Colorimetric Method for the Determination of Sugars and Related Substances. Nature. 1956;**168**:167

[45] FAO. FAO codex alimentarius nitrogen conversion factor for Soy Protein - Joint FAO/WHO Food Standards Programme codex committee on methods of analysis sampling. The Joint FAO/WHO Expert Meetings on Nutrition (JEMNU); 2016.

[46] Folch J, Lees M, Sloane Stanley GH. Method for the isolation and purification of total lipides from animal tissues. Journal of Biological Chemistry. 1957;**226**:497–509

[47] Conde TA, Neves BF, Couto D, Melo T, Neves B, Costa M, et al. Microalgae as sustainable bio-factories of healthy lipids: Evaluating Fatty Acid content and antioxidant activity. Marine Drugs. 2021;**19**:1–20

[48] Rodriguez De Marco E, Steffolani ME, Martinez CS, Leon AE. Effects of spirulina biomass on the technological and nutritional quality of bread wheat pasta. Food Science and Technology. 2014;**58**:102–108

[49] Aouir A, Amiali M, Bitam A, Benchabane A, Raghavan VG. Comparison of the Biochemical Composition of Different Arthrospira platensis Strains from Algeria, Chad and the USA. Journal of Food Measurement and Characterization. 2017;**11**:913–923. DOI: 10.1007/s11694-016-9463-4

[50] Ogbonda KH, Aminigo RE, Abu GO. Influence of temperature and pH on biomass production and protein biosynthesis in a putative Spirulina sp. Bioresource Technology. 2007;**98**: 2207–2211

[51] Sharoba AM. Nutritional value of Spirulina and its use in the preparation of some complementary baby food formulas. Journal of Food and Dairy Sciences. 2014;**5**:517–538

[52] Senila L, Kovacs E, Roman C. Chemical Characterization, Lipid Profile, and volatile compounds in Chlorella sp. and Spirulina platensis: A promising feedstock for various applications. Molecules. 2025;**30**:1499

[53] Rahim A, Çakir C, Ozturk M, Sahin B, Soulaimani A, Sibaoueih M, et al. Chemical characterization and nutritional value of Spirulina platensis cultivated in natural conditions of Chichaoua region (Morocco). South African Journal of Botany. 2021;**141**:235–242

[54] Burnap RL, Hageman M, Kaplan A. Regulation of CO2 concentrating

mechanism in Cyanobacteria. Life. 2015;**5**:348–371

[55] Herrero A, Muro-Pastor AM, Valladares A, Flores E. Cellular differentiation and the NtcA transcription factor in filamentous cyanobacteria. FEMS Microbiology Reviews. 2004;**28**:469–487

[56] Chellamuthu VR, Alva V, Forchhammer K. From cyanobacteria to plants: Conservation of PII functions during plastid evolution. Planta. 2013;**237**:451–462

[57] Giordano M. Homeostasis: An underestimated focal point of ecology and evolution. Plant Science. 2013;**211**:92–101

[58] Du J, Förster B, Rourke L, Howitt SM, Price GD. Characterisation of cyanobacterial bicarbonate transporters in E. coli shows that SbtA homologs are functional in this heterologous expression system. PLoS One. 2014;**9**(12):e0115905. DOI: 10.1371/journal.pone.0115905

[59] Klahn S, Bolay P, Wright PR, Atilho RM, Brewer KI, Hagemann M, et al. A glutamine riboswitch is a key element for the regulation of glutamine synthetase in cyanobacteria. Nucleic Acids Research. 2018;**46**:10082–10094

[60] Selim KA, Haese F, Hartmann MD, Hagemann M, Forchhammer K. PII-like signaling protein SbtB links cAMP sensing with cyanobacterial inorganic carbon response. Pnas. 2018;**115**(21): E48661–E4869. DOI: 10.1073/pnas.1803790115

[61] Demirbas A. Biomass resource facilities and biomass conversion processing for fuels and chemicals. Energy Conversion and Management. 2001;**42**:1357–1378

[62] Shastri AA, Morgan JA. Flux balance analysis of photoautotrophic metabolism. Biotechnology Progress. 2005;**21**:1617–1626

[63] Cai Y, Zhai L, Fang X, Wu K, Liu Y, Cui X, et al. Effects of C/N ratio on the growth and protein accumulation of heterotrophic Chlorella in broken rice hydrolysate. Biotechnology for Biofuels and Bioproducts. 2022;**15**:102

[64] Yousefi R, Mottaghi A, Saidpour A. Spirulina platensis effectively ameliorates anthropometric measurements and obesity-related metabolic disorders in obese or overweight healthy individuals: A randomized controlled trial. Complementary Therapies in Medicine. 2018;**40**:106–112

[65] Khafaga AF, El-Sayed YS. Spirulina ameliorates methotrexate hepatotoxicity via antioxidant, immune stimulation, and proinflammatory cytokines and apoptotic proteins modulation. Life Sciences. 2018;**196**: 9–17

[66] Zeinalian R, Farhangi MA, Shariat A, Saghafi-Asl M. The effects of Spirulina Platensis on anthropometric indices, appetite, lipid profile and serum vascular endothelial growth factor (VEGF) in obese individuals: A randomized double blinded placebo controlled trial. BMC Complementary and Alternative Medicine. 2017;**17**:225

[67] CY C, YC C, Hc H, CC H, Lee WL, Chang JS. Engineering strategies for enhancing the production of eicosapentaenoic acid (EPA) from an isolated microalga Nannochloropsis

oceanica CY2. Bioresource Technology. 2013;**147**:160–167

[68] Mariamenatu AH, Abdu EM. Overconsumption of Omega-6 Polyunsaturated Fatty Acids (PUFAs) versus deficiency of Omega-3 PUFAs in modern-day diets: The disturbing factor for their "Balanced Antagonistic Metabolic Functions" in the human body. Journal of Lipids. 2021;**2021**:1–15

[69] Dabrowska NK, Marcinkowski K, Mazur A, Mazur S, Madera M, Strus K, et al. Spirulina maxima supplementation: Benefits and limitations – Results of latest studies. Journal of Pre-Clinical and Clinical Research. 2024;**18**:195–200

[70] Uzlasir T, Selli S, Kelebek H. Spirulina platensis and Phaeodactylum tricornutum as sustainable sources of bioactive compounds: Health implications and applications in the food industry. Future Postharvest and Food. 2024;**1**:34–46

[71] Wu Z, Xia F, Wang W, Zhang K, Fan M, Lin R. The global burden of disease attributable tohigh body mass index in 204 countries andterritories from 1990 to 2021 with projectionsto 2050: An analysis of the Global Burden of Disease Study 2021. European Journal of Heart Failure. 2025;**27**:354–365

[72] Ahmed SK, RA M. Obesity: Prevalence, causes, consequences, management, preventive strategies and future research directions. Metabolism Open. 2025;**27**:100375

[73] Hewelt-Belka W, Kot-Wasik A, Tamagnini P, Oliveira P. Untargeted Lipidomics analysis of the Cyanobacterium Synechocystis sp. PCC 6803: Lipid composition variation in response to alternative cultivation setups and to gene deletion. International Journal of Molecular Sciences. 2020;**23**:8883

[74] Sheng J, Vannela R, Rittmann BE. Evaluation of methods to extract and quantify lipids from Synechocystis PCC 6803. Bioresource Technology. 2011;**102**:9–17

[75] V. UTL, Southgate DAT. Coronary Heart disease: Seven dietary factors. The Lancet. 1991;**338**:985–992

[76] Naeini Z, Toupchian O, Vatannejad A, Sotoudeh G, Teimouri M, Ghorbani M, et al. Effects of DHA-enriched fish oil on gene expression levels of p53 and NF-κB and PPAR-γ activity in PBMCs of patients with T2DM: A randomized, double-blind, clinical trial. Nutrition, Metabolism and Cardiovascular Diseases. 2020;**30**:441–447

[77] Chen J, Liu H. Nutritional indices for assessing Fatty Acids: A Mini-Review. International Journal of Molecular Sciences. 2020;**21**:5695

[78] Minerals KM. PUFAs and antioxidant properties of some tropical seaweeds from Saurashtra coast of India. Journal of Applied Phycology. 2011;**23**:797–810

[79] Rincón-Cervera MA, González-Barriga V, Romero J, Rojas R, López-Arana S. Quantification and distribution of Omega-3 Fatty acids in South Pacific Fish and Shellfish Species. Foods. 2020;**9**:233

[80] Fernandes CE, Vasconcelos MAS, RIbeiro MA, Sarubbo LA, Andrade SAC, Filho ABM. Nutritional and lipid profiles in marine fish species from Brazil. Food Chemistry. 2014;**160**: 67–71

[81] Mir NA, Tyagi PK, Tyagi AKBPK, Mandal AB, Kumar F, Sharma D, et al. Inclusion of Flaxseed, Broken Rice, and Distillers Dried Grains with Solubles (DDGS) in Broiler Chicken Ration Alters the Fatty Acid Profile, oxidative stability, and other functional properties of meat. European Journal of Lipid Science and Technology. 2018;**120**:1700470

[82] Wood JD, Richardson RI, Nute GR, V. FA, Campo MM, Kasapidou E, et al. Effects of fatty acids on meat quality: A review. Meat Science. 2004;**66**:21–32

[83] Simopoulos AP. he Importance of the Omega-6/Omega-3 Fatty Acid Ratio in Cardiovascular disease and other chronic disease. Experimental Biology and Medicine. 2008;**233**: 674–688

[84] Puchau B, Ochoa MC, Zulet MA, Marti A, Martínez JA. Dietary total antioxidant capacity and obesity in children and adolescents. International Journal of Food Sciences and Nutrition. 2010;**61**:713–721

[85] Park WS, Kim HJ, Li M, Lim DH, Kim J, Kwak SS, et al. Two classes of pigments, Carotenoids and C-Phycocyanin, in Spirulina powder and their antioxidant activities. Molecules. 2018;**23**:2065

[86] Freitas S, Silva NG, Sousa ML, RIbeiro T, Rosa F, Leão PN, et al. Chlorophyll derivatives from Marine Cyanobacteria with Lipid-reducing activities. Marine Drugs. 2019;**17**:229

[87] Ferrruzzi MG, Blakeslee J. Digestion, absorption, and cancer preventative activity of dietary chlorophyll derivatives. Nutrition Research. 2007;**27**:1–12

[88] Tufail T, Bader Ul Ain H, Noreen S, Ikram A, Arshad MT, Abdullahi MA. Nutritional benefits of lycopene and beta-carotene: A comprehensive overview. Food Science & Nutrition. 2024;**12**:8715–8741

[89] Gebregziabher BS, Gebremeskel H, Debesa B, Ayalneh D, Mitiku T, Wendwessen T, et al. Carotenoids: Dietary sources, health functions, biofortification, marketing trend and affecting factors – A review. Journal of Agricultural and Food Research. 2023;**14**:100834

[90] Krinsky NI, Johnson EJ. Carotenoid actions and their relation to health and disease. Molecular Aspects of Medicine. 2005;**26**:459–516

[91] Scripsema NK, Hu DN, Rosen RB. Lutein, zeaxanthin, and meso-zeaxanthin in the clinical management of eye disease. Journal of Ophthalmology. 2015;**2015**:865179

[92] Patel A, Mishra S, Pawar R, Ghosh PK. Purification and characterization of C-Phycocyanin from cyanobacterial species of marine and freshwater habitat. Protein Expression and Purification. 2005;**40**:248–255

[93] Pan-Utai W, Iamtham S. Extraction, purification and antioxidant activity of phycobiliprotein from Arthrospira platensis. Process Biochemistry. 2019;**82**:189–198

[94] Gabr GA, El-Sayed SM, Hikal MS. Antioxidant Activities of Phycocyanin: A Bioactive Compound from Spirulina platensis. Journal of Pharmaceutical Research International. 2020;**32**:73–85

[95] Benedetti S, Benvenuti F, Pagliarini S, Francogli S, Scoglio S, Canestrani F. Antioxidant properties of a novel phycocyanin extract from the blue-green alga Aphanizomenon flos-aquae. Life Sciences. 2004;**75**: 2353–2362

[96] Cano-Europa E, Ortiz-Butrón R, Gallardo-Casas CA, Blas-Valdivia V, Pineda-Reynoso M, Olvera-Ramírez R, et al. Phycobiliproteins from Pseudanabaena tenuis rich in c-phycoerythrin protect against HgCl2-caused oxidative stress and cellular damage in the kidney. Journal of Applied Phycology. 2010;**22**:495–501

[97] Gan F, Zhang S, Rockwell NC, Martin SS, Lagarias JC, Bryant DA. Extensive remodeling of a cyanobacterial photosynthetic apparatus in far-red light. Science. 2014;**345**:1312–1317

[98] Sathuvan M, Thangam R, Venkateshbabu G, Cheong KL, Kang H, Liu Y. Single-step purified R-phycoerythrin transmits cellular imaging functionalities in vitro. International Journal of Biological Macromolecules. 2022;**19**:563–570

[99] Shahidi F, Ambigaipalan P. Phenolics and polyphenolics in foods, beverages and spices: Antioxidant activity and health effects – A review. Journal of Functional Foods. 2015;**18**:820–897

[100] Osawa T, Narasimhan R, Kawakishi S, Namdci M, Tashiro T. Antioxidative defense systems in rice hull against damage caused by oxygen radicals. Agricultural and Biological Chemistry. 1985;**49**:3085–3087

[101] El.-Baky HHA, Baz FK, El-Baroty GS. Production of phenolic compounds from Spirulina maxima microalgae and its protective effects in vitro toward hepatotoxicity model. African Journal of Pharmacy and Pharmacology. 2009;**3**:133–139

[102] Okburan G, Gezer C. Carbohydrates as nutritional components for health and longevity. In: Rattan SIS, Kaur G, editors. Nutrition, Food and Diet in Ageing and Longevity. Healthy Ageing and Longevity. Vol. **14**. New York: Springer; 2021. p. 1152

Chapter 7

Limno-Ecological Changes That Influenced Peridinium Decline and Cyanobacteria Enhancement in Lake Kinneret, Israel

Moshe Gophen

Abstract

Lake Kinneret is the only natural freshwater lake in Israel. The lake and its headwaters are utilized for approximately 26% of the national water budget, including supply, commercial fishery, recreation, and tourism. The water quality of Lake Kinneret is, therefore, a national concern. The area of the Kinneret drainage basin is 2,730 km^2 of which the entire Hula Valley is about 200 km^2. Three major headwater rivers flow southward and jointly create the Jordan River, which then flows into Lake Kinneret. The Jordan River contributes approximately 63% of the Kinneret water budget and more than 50% of the total external nutrient inputs, which are critical for algal growth. In the 1950s, the wetland and old Lake Hula were drained, and the land was converted from a natural ecosystem habitat to agricultural development. Before drainage, the Hula wetlands area was densely covered by submerged and emergent aquatic vegetation, and its water cover was mostly anoxic, containing ammonium, with no nitrates detected. Until the late 1990s, the Kinneret phytoplankton communities were predominantly dominated by Pyrrhophyta (*Peridinium sp.*), but later, cyanobacteria took over, leading to deteriorating water quality. The historical record of the N_2-fixing toxic cyanobacterium, *Aphanizomenon ovalisporum,* blooming signaled a warning of water quality deterioration. Evaluation of long-term data records of lake and watershed nutrient dynamics investigates the upheaval of phytoplankton assemblage composition change in the lake.

Keywords: Kinneret, Hula, N-insufficiency, P-sufficiency, peridinium, cyanobacteria

1. Introduction

A 10-year (1969–1979) non-modulated dynamical comprehensive structure of the Kinneret ecosystem was documented, entitled "Model A" [1]. Twenty-five years later, a modulated evaluation of a 32-year (1969–2001) dynamical comprehensive structure Lake Kinneret database, with a scoped focus on the role of zooplankton, was documented, entitled "Model B" [2]. Although there is a partial dissimilarity between the two documents, there is a conceptual overlap between them. The usage of metabolic and geochemical ecological coefficients measures interaction rates between ecosystem compartments. Model B includes temperature distribution partitioning, and Model A emphasizes the role of fish [3–6]. Model B illustrates the potentially useful information to complement field data collection with respect to the role of zooplankton in nutrient cycles. Model B is coupled to physical modulation, whilst Model A relies solely on food web dynamics as thermally affected. Model B quantifies the role of zooplankton in the distribution and flux of Carbon, whilst Model A illustrates the carbon flow pattern throughout all ecosystem compartments. The two models do not illustrate predictive instructions for management design aimed at preventing deterioration or improving water quality in Lake Kinneret, although fishery management through the removal of unwanted zooplankton consumer fishes is implied in Model A [1, 3, 4, 7, 8]. The impact of nutrient dynamics and composition on water quality is not concluded in either model. The objective of this paper is to evaluate the temporal dynamics of internal and external nutrient input with respect to climate conditions. The consequences on the structure of phytoplankton communities and the process of Pyrrhophyte decline and enhancement of cyanobacteria are analyzed.

During the summer of 1994, a bloom of the harmful cyanobacterium (HFCB), the predictable *Aphanizomenon ovalisporum* (AO), was recorded in Lake Kinneret for the first time since the early 1960s [3, 9–12]. Therefore, the Lake Kinneret ecosystem could be classified, like many other lakes, as a potential HFCB ecosystem, where insufficient nitrogen is accompanied by phosphorus sufficiency, causing a decline in the N/P mass ratio [9–16]. The documented ecological parameters that enabled the first historical outbreak of the N_2-fixing cyanobacterium, AO, were widely studied [10–13, 17–18]. Despite several studies relevant to eco-limnological changes in the Lake Kinneret ecosystem [3–6, 8, 9, 12, 20, 21], a comprehensive ecological retrospective and future prediction of the entire ecosystem, including the watershed and the lake as an integrated system, along with consequent management design, is missing. The objective of this paper is to fill this knowledge gap from a management perspective.

Lake Kinneret and its watershed, including the Hula Valley, are located in northern Israel's Syrian–African Great Rift Valley. Lake Kinneret is the only natural freshwater lake in Israel. Between 1970 and 2010, an annual average of 336 mcm (10^6 m^3) of Kinneret water was withdrawn from the lake and supplied to the south for domestic and agricultural irrigation usage. The water quality of Lake Kinneret is, therefore, a national concern. The total national water supply is 2.11 bcm ($10^9 m^3$), of which 0.55 bcm comes from the Kinneret–Jordan hydrological system. The area of the Kinneret drainage basin is 2,730 km^2 and is located mostly to the north of the lake, with the total Hula Valley covering about 200 km^2. Three major headwater rivers (Hatzbani, Banyas, and Dan) flow from the Hermon Mountain region in the north, join together, and form the Jordan River, which crosses the Hula Valley and flows into Lake Kinneret. Until 1957, the Jordan crossed the Hula Valley through three branches (tributaries) flowing into the old Lake Hula. From Lake Hula, at an altitude of 61–68 masl (meters above mean sea level), the Jordan River flows downstream into Lake Kinneret (mean water level 211 mbsl) for a distance of approximately 15 km. The altitude difference between the top of Mount Hermon (2,814 masl) and Lake Kinneret's water level (mean 211 mbsl) along a spatial distance of 70 km results in a mean slope of 4.3%, creating a rather strong erosive force. The Jordan River contributes approximately 63% of the Kinneret water budget and more than 50% of the total external nutrient inputs. Until

1957, the Hula Valley land was covered by old Lake Hula (1.5 m mean depth; 13 km^2 water surface) and 3,500 ha of swampy wetlands. The wetland area was covered by water in the winter and partly covered in the summer. To the north of the swamps was an area (3,200 ha) where water table levels were high in winter, thus making agricultural cultivation impossible. During the summer, when underground water levels declined, this 3,200-ha land surface was efficiently cultivated. The wetland area was densely covered by submerged and emergent aquatic vegetation, and the water cover was mostly anoxic, containing ammonium, with no nitrate detected [22].

The objective of this study is to reveal the ecological background that created the current development. The unique bloom outbreak of AO in the summer of 1994, the depleted biomass density of the routinely formed *Peridinium* bloom, and the enhancement of cyanobacteria instead were widely investigated and documented. The impact of external and internal nutrient inputs that induced those changes is the focused narrative of the present study. The skeletal concept of this paper is not the highly documented information about the resulting conditions, but the background induction of it. Lake conditions maintained during the occurrence of the AO outbreak were intensively investigated. On the other hand, the tentative factors of the environmental background that induced the enhancement of cyanobacteria were not thoroughly studied and are the focus of this paper.

2. Materials and methods

2.1 Data sources

The data about nutrient concentrations and phytoplankton density were taken from the Kinneret Limnological Laboratory Data Base and Annual (1969–2001) reports [23] (M. Schlichter). Annual (1969–2001) and monthly data of rainfall capacities measured at the Dafna (northern part of the Hula Valley) meteorological station were obtained from the Israeli Meteorological Service (courtesy of M. Peres). Nitrogen and phosphorus concentrations in River Jordan (Huri Station) were taken from Annual (1967–2001) reports.

2.2 Kjeldahl (organic nitrogen plus ammonia)

Total-unfiltered sample; dissolved-filtered sample; digestion – H_2SO_4; distillation – NaOH; titration – ammonia.

Total nitrogen (TN), nitrate (NO_3): Diphenylamine sulfuric acid chromogen – NAS method.

Ammonia (NH4): Phenol-hypochlorite method.

Total phosphorus (TP) and total dissolved phosphorus (TDP): Persulfate oxidation/ascorbic acid method.

2.3 Statistical evaluation

Statistical evaluation was conducted using the software STATA 17.0 – Standard Edition, Statistics and Data Science, Copyright 1985–2021 StataCorp LLC, StataCorp, 4905 Lakeway Drive, 800-STATA-PC. The Stata license is a single-user perpetual license, serial number: 401706315938, licensed to Moshe Gophen, Migal. Three statistical methods were utilized: Linear and fractional polynomial (with 95% confidence limits), regression (prediction), and biplot PCA multivariate analysis.

3. Results

3.1 Monthly distribution of nutrient concentrations in Jordan water

The monthly and temporal distribution of nitrogen, phosphorus, and the TN/TP ratio in the Jordan water (Huri Station) are shown in the following figures and tables.

Results given in **Figures 1–3** confirm the *Peridinium* response to TN enrichment by division rate and biomass density enhancement during winter, spring, and early summer, and biomass decline during mid-summer and fall in response to TN diminishment. The TN periodical increase starts earlier than the TP elevation. Previous studies [1, 13, 17, 21, 24–26] confirmed *Peridinium* TP-mediated biomass transportation factor from bottom sediments into the epilimnion, whereas TN epilimnetic sources are mostly external and partly derived from ammonium oxidation. Consequently, the fall-winter increase of the TN load in the epilimnion initiates earlier than that of TP when the destratification process progresses. The *Peridinium* bloom intensity is dependent on the dynamics (rate and range) of nitrogen import from the catchment through headwater flows. The higher the nitrogen, the denser the *Peridinium* biomass. However, despite the pattern similarity of nitrogen and phosphorus external import, the *Peridinium* bloom biomass density does not depend on epilimnetic phosphorus availability. Moreover, the *Peridinium* bloom is initiated by the excystation process of cysts within bottom sediments, transforming into a movable form equipped with a flagellum. These vegetative mobile cells are phosphorus-rich, having incorporated phosphorus from the sediment substrate. The bloom development is promoted by an intensive division

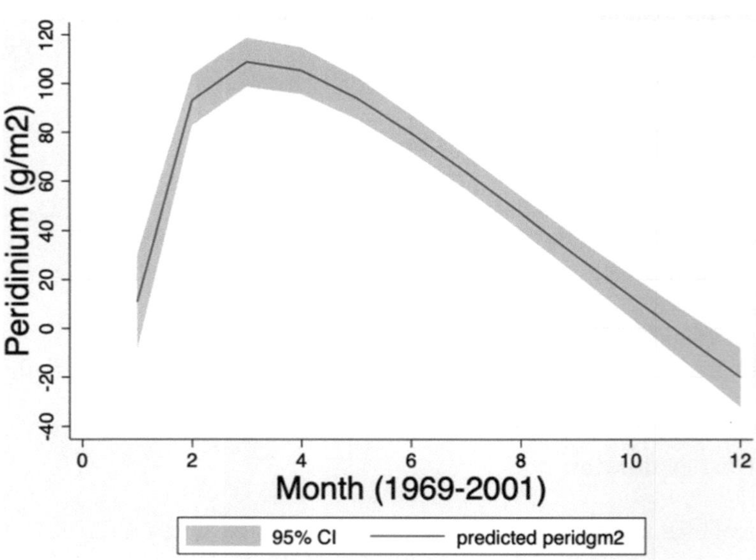

Figure 1.
Fractional polynomial (with 95% confidence limit) correlative regression plot between monthly means of Pyrrhophyta (Peridinium spp.) biomass (g(ww)/m²) and month in Lake Kinneret during 1969–2001.

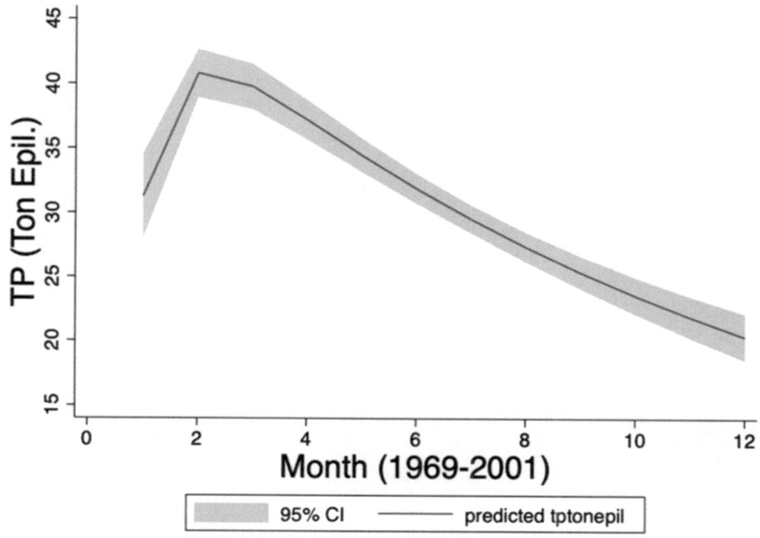

Figure 2.
Fractional polynomial (with 95% confidence limit) correlative regression plot between monthly means of total phosphorus (TP) epilimnetic load (tons) and month in Lake Kinneret during 1969–2001.

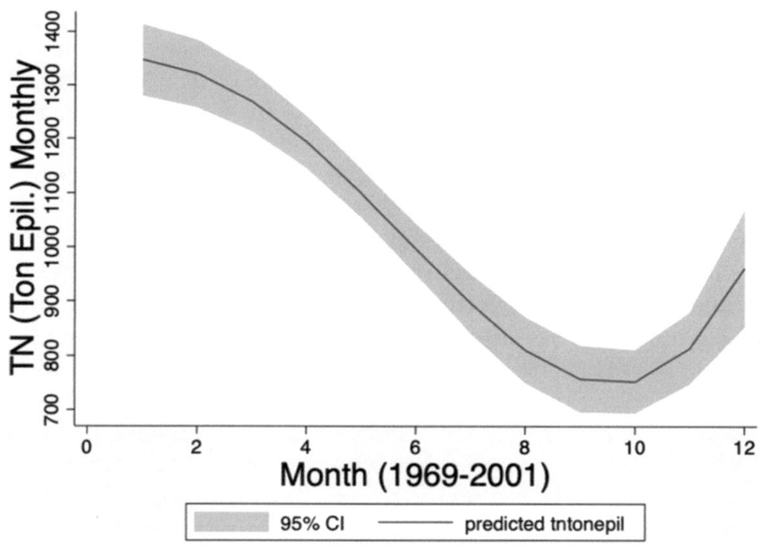

Figure 3.
Fractional polynomial (with 95% confidence limit) correlative regression plot between monthly means of total nitrogen (TN) epilimnetic load (tons) and month in Lake Kinneret during 1969–2001.

rate, creating a factor of epilimnetic phosphorus enhancement. *Peridinium* cells' natural mortality, degradation, and decomposition induce the release of dissolved organic phosphorus (DIN) to the epilimnion. The bloom inset starts in late winter and offsets occur during early summer (June) [1, 27]. This phenomenon sequence of a winter-heavy *Peridinium* inset bloom, terminated by early summer with intensive

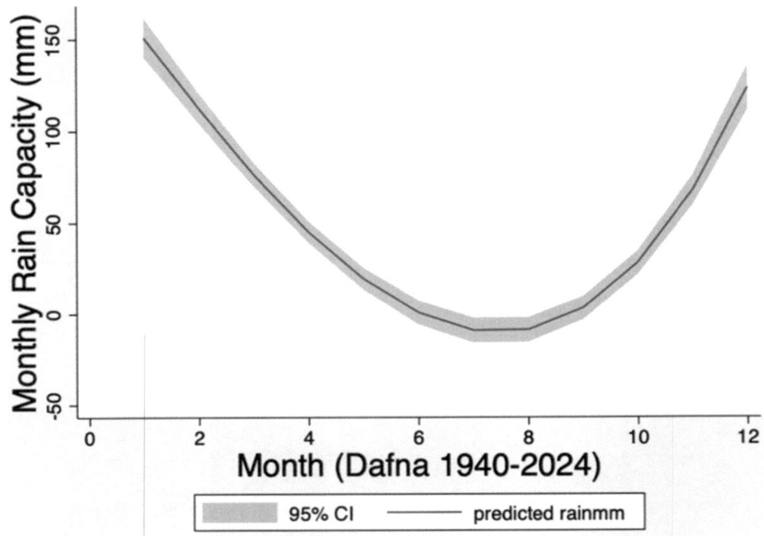

Figure 4.
Fractional polynomial (with 95% confidence limit) correlative regression plot between monthly means (1940–2024) of rainfall capacity (Dafna) and month.

cyst formation (cystation), followed next year by epilimnetic high densities of vegetative bloom inset, and consequently bloom offset collapse and high DIN concentration in early summer, enabled the outbreak of AO. Moreover, these complex interactions between climate conditions, nutrient dynamics, and cyanobacteria in Lake Kinneret are a repeatable situation. These conditions are suitable for N_2-fixer cyanobacterium blooming: Low nitrogen, sufficient phosphorus, and diminished competitor *Peridinium* biomass. What else is conceivable for the HFCB outbreak? A reasonable inoculum.

Results shown in **Figures 3** and **4** indicate the correlation between rainfall, river discharges, and the epilimnetic nitrogen load, highlighting the impact of externally sourced nitrogen input. River discharge plays a role as a transporter in the Kinneret ecosystem, with the transport load range being significantly influenced by rainfall capacity and flow rate. Typically, high loads occur in winter and low loads in summer. However, the transport range depends on the availability of nutrient sources. Erosion caused by water flow friction on the landscape's rocky infrastructure is greatly affected by the natural discharge range. Nonetheless, the contribution of anthropogenic resources, such as human and domestic dairy sewage (raw and treated) or fishpond effluents, is independent of the natural rainfall-discharge regime. Before the Hula drainage, nitrogen input loads into Lake Kinneret, mainly as ammonium, were significantly increased by *Peridinium.* After the Hula drainage, ammonium influx from the wetlands was replaced by peat soil components and nitrate, which has a lower affinity for *Peridinium*. However, this shift is complicated by sewage and fishpond effluents. Since the late 1980s, following the removal of domestic and dairy sewage and the conversion of fishponds to field crops, organic nitrogen input was reduced [28], leading to a period of N-limitation and a decline in the N/P ratio. These conditions favored cyanobacteria over *Peridinium*, creating a less suitable environment for their growth.

3.2 Temporal (1970–2001) distribution of nutrient concentrations in Jordan water

Results shown in **Figure 5** represent the correlation between TN and TP epilimnetic load sizes ordered from small to large. **Figure 5** emphasizes the impact of external and internal, different or similar properties, on TN and TP. Nevertheless,

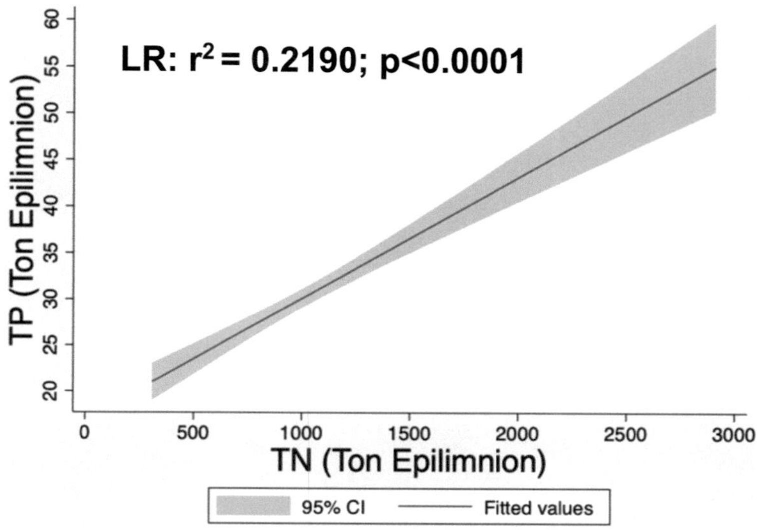

Figure 5.
Linear regression (with 95% confidence limit) plot between monthly means of TN and TP epilimnetic loads during 1969–2001.

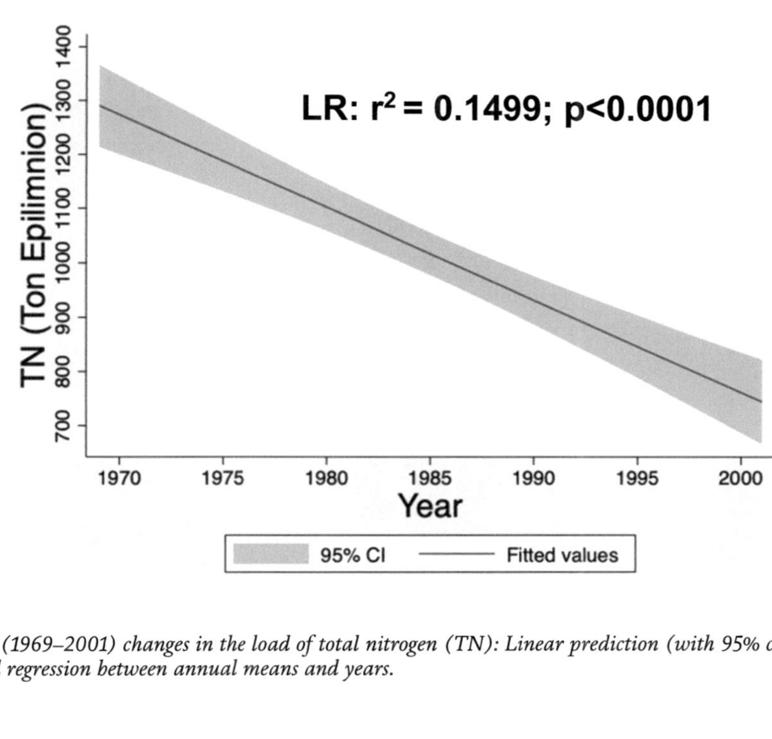

Figure 6.
Temporal (1969–2001) changes in the load of total nitrogen (TN): Linear prediction (with 95% confidence limit) and regression between annual means and years.

the temporal sequence record of the mass ratio between those nutrients resulted in a declining trend. **Figure 5** represents, probably, a misleading, distorted conclusion, which is corrected by the mass ratio time series record. The amendment is an ecological improvement that is beneficial for water quality management design.

Results given in **Figures 6–8** indicate that the nitrogen decline in the Kinneret epilimnion during 1969–2001 contributes a dominant impact on the TN/TP ratio changes.

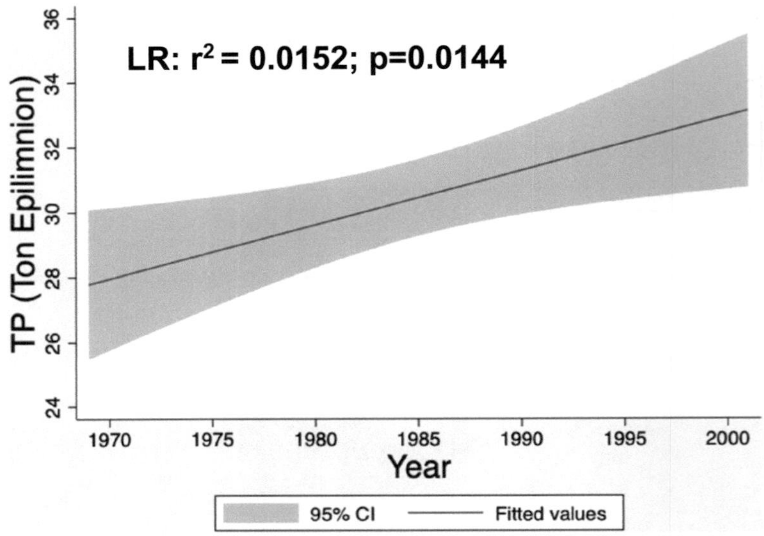

Figure 7.
Temporal (1969–2001) changes in the load of total phosphorus (TP): Linear prediction (with 95% confidence limit) and regression between annual means and years.

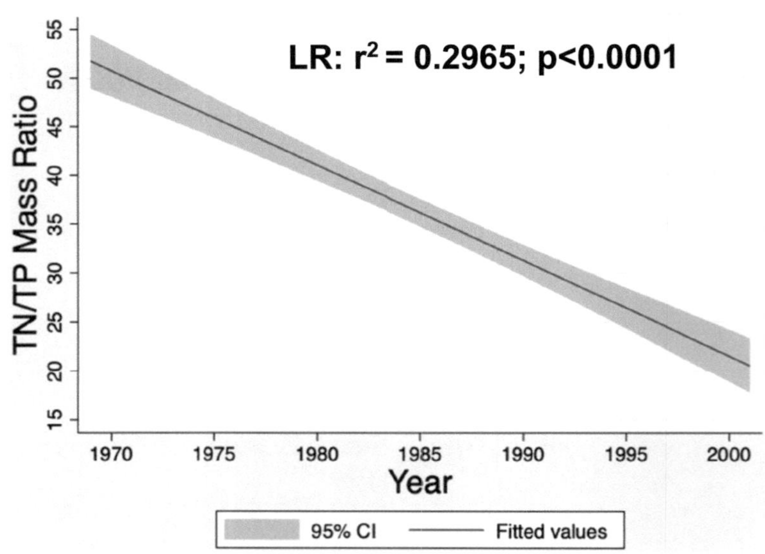

Figure 8.
Temporal (1969–2001) changes in annual (1969–2001) means of epilimnetic TN/TP, loads of TN and TP: Linear prediction (with 95% confidence limit) and regression between ratios and years.

3.3 Lake nutrients-phytoplankton relations

The data given in **Table 1** indicate the decline of algal biomass with an increasing TN/TP mass ratio. An increasing TN/TP ratio is most probable where the comprehensive ecosystem nutrient composition structure represents algal phosphorus limitation.

Results given in **Table 2** indicate phytoplankton enhancement, as well as epilimnetic TP elevation and a decline in TN loads.

The monthly data given in **Table 3** confirm a significant summer-fall decline in phytoplankton biomass and nutrient loads (except Chlorophyta), while Cyanophyta and Chlorophyta show enhancement.

Algal group	r^2	p
Total Phytoplankton	0.0416	<0.0001
Non-Pyrrhophyta	0.0342	0.0003
Cyanophyta	0.0141	0.0189
Bacillariophyta	0.0148	0.0161
Chlorophyta	0.0283	0.0008
Pyrrhophyta	0.0191	0.0063

Table 1.
Linear regression parameters (r^2, p) of the correlations between monthly means of TN/TP ratio values and phytoplankton groups (Chlorophyta, Bacillariophyta, Cyanophyta, Pyrrhophyta, total phytoplankton, and non-Pyrrhophyta) mean epilimnetic biomass (g(ww)/m^2) in the Kinneret epilimnion during 1969–2001.

Nutrient/algal group	Trend	Significance	r^2	p
TP	IN	S	0.0171	$p = 0.0096$
TN	DE	S	0.1429	$p < 0.0001$
Cyanophyta	IN	S	0.0638	$p < 0.0001$
Bacillariophyta	IN	S	0.0553	$p < 0.0001$
Chlorophyta	IN	S	0.0965	$p < 0.0001$
Pyrrhophyta	DE	NS	0.0025	$p = 0.9040$

Table 2.
Linear regression parameters (r^2, p) (S = significant; NS = not significant) of the correlations between temporal changes (IN = increase; DE = decline) (annual means) from 1969–2001 of nutrients (TN, TP; tons) and phytoplankton groups (Chlorophyta, Bacillariophyta, Cyanophyta, Pyrrhophyta) biomass (g(ww)/m2m^2) in the Lake Kinneret epilimnion. The annual means of total phytoplankton, which is dominated by Pyrrhophyta, showed a low-level significant increase ($r^2 = 0.0170$; $p = 0.0098$).

Nutrient and algal group	Trend	Significance	r^2	p
TP	DE	S	0.2319	$p < 0.0001$
TN	DE	S	0.1770	$p < 0.0001$
Total Phytoplankton	DE	S	0.2088	$p < 0.0001$
Non-Pyrrhophyta	DE	S	0.0317	$p = 0.0004$
Cyanophyta	IN	S	0.0241	p-0.0021
Bacillariophyta	DE	S	0.0630	$p < 0.0001$
Chlorophyta	IN	NS	0.0006	$p = 0.6267$
Pyrrhophyta	DE	S	0.1489	$p < 0.0001$

Table 3.
Linear regression parameters (r^2, p) (S = significant; NS = not significant) of the correlations between monthly changes (IN = increase; DE = decline) (monthly means) during 1969–2001 of nutrients (TN, TP; tons) and phytoplankton group biomass (g(ww)/m^2): Total phytoplankton, non-pyrrhophyta, chlorophyta, bacillariophyta, cyanophyta, and pyrrhophyta in the Lake Kinneret epilimnion.

3.4 Periodical changes of nutrient concentrations in Jordan waters

3.5 The nutrient correlation with *Peridinium* and Cyanophyta

The results given in **Figure 9** confirm a significant correlation between TN load and *Peridinium* biomass in the epilimnion of Lake Kinneret. The definition of external and internal-supported nitrogen availability as a limiting factor for *Peridinium* algal growth is therefore confirmed.

The results given in **Figure 10** confirm a significant correlation between TP load and *Peridinium* biomass in the epilimnion of Lake Kinneret. It is suggested that the TP load increase in the Kinneret epilimnion resulted from internal P-mediated proliferation of *Peridinium* cells. In other words, phosphorus is transported through

Period / Nutrient	1970–1985	1986–2000	2001–2018
NH_4	0.138	0.089	0.058
NO_3	1.619	1.932	2.021
N-org.	1.030	0.550	0.381
Kjeldhal total	1.168	0.640	0.434
Kjeldhal dissolved	0.577	0.332	0.232
TN	2.774	2.565	2.44
TP	0.236	0.176	0.117
TDP	0.045	0.040	0.033
mcm	467	427	355
TN/TP	11.75	14.57	20.85

Table 4.
Periodical (1970–1985; 1986–2000; 2001–2018) means of nutrient concentrations (ppm): Ammonium (NH_4), nitrate (NO_3), organic nitrogen (n-org.), Kjeldahl total, Kjeldahl dissolved, total nitrogen (TN), total phosphorus (TP), and TN/TP ratio (ppm/ppm) in the Jordan River (Huri Station).

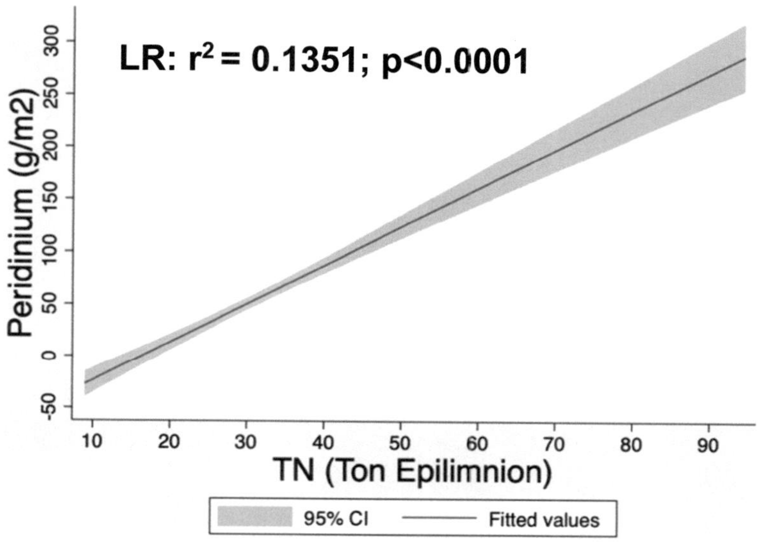

Figure 9.
Linear prediction (with 95% confidence limit) correlative regression between monthly means of Peridinium spp. biomass (g(ww)/m^2) and epilimnetic TN load (tons) in Lake Kinneret during 1969–2001.

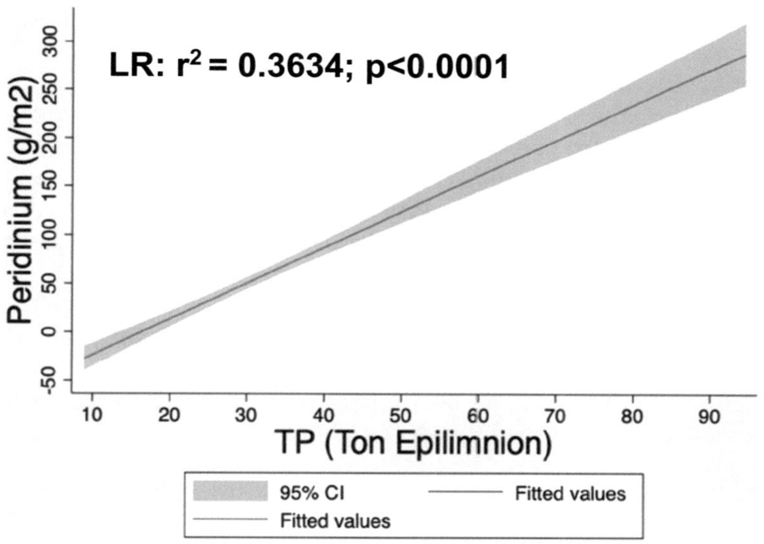

Figure 10.
Linear prediction (with 95% confidence limit) correlative regression between monthly means of Peridinium *spp. biomass (g(ww)/m^2) and epilimnetic TP load (tons) in Lake Kinneret during 1969–2001.*

vegetative *Peridinium* cyst germination, migrating upwards into the light-penetrated, warm, and oxidized epilimnion layer. Among the four major phytoplankton algal groups – Chlorophyta, Bacillariophyta (diatoms), Pyrrhophyta (*Peridinium spp.),* and Cyanophyta (blue-green) – only blue-greens presented significant temporal (1969–2001) changes in biomass density. Evaluated statistical correlations between grouped non-Pyrrhophyte phytoplankters and TN and TP were calculated: Non-Pyrrhophyte versus TN was low-level significant (r^2 = 0.0201; p = 0.0058), whilst a slightly higher

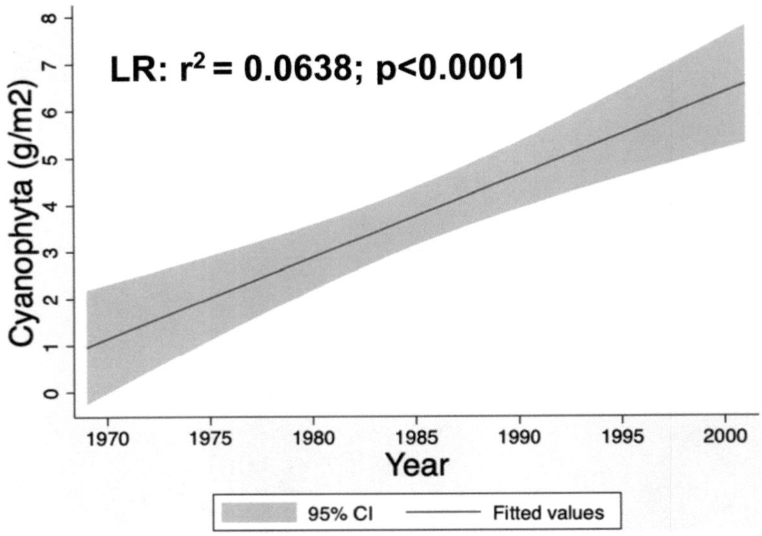

Figure 11.
Temporal (1969–2001) changes tested as linear regression (with 95% confidence limit) between annual mean biomass density (g(ww)/m^2) of cyanophyta biomass and years in Lake Kinneret.

level of significance was observed with TP: r^2 = 0.0359 and p = 0.0009. Significant correlation was indicated between TN and cyanophytes and *Peridinium*. Consequently, reduced availability of TN was followed by *Peridinium* decline, whilst Cyanophyta enhancement. It is suggested that the low-significant positive correlation found between non-Pyrrhophyta and nutrient (TN, TP) availabilities indicates algal growth enhancement, confounding the top-down cascading effect of grazing depression through the removal of unwanted zooplankton predator fish management. Nevertheless, *Peridinium* consumption by zooplankton (except for large-bodied rare rotifers, *Asplanchna spp.*) was not recorded. **Figures 11, 14.**

3.6 PCA analysis of nutrients and phytoplankton predominance

The inquiry into this issue was not emphasized by Kinneret investigators and ecosystem model constructors. In the early 1980s, scientific discoveries about

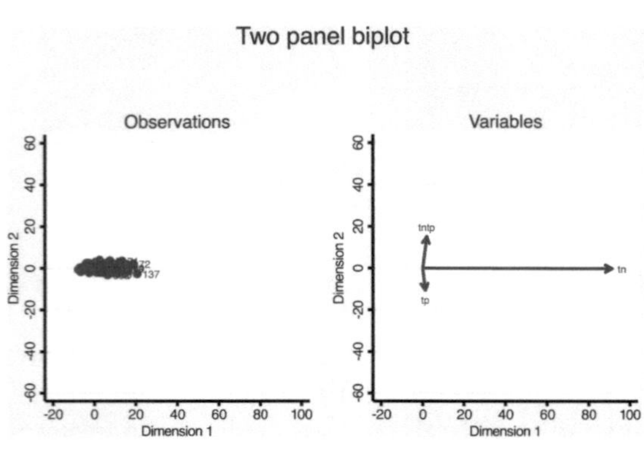

Table of Biplot Coordinates

Variables	Dimension	Dimension 2
TP	1.1987.	-10.5568
TN	91.4478	-0.1839
TN/TP	1.9393	15.1959

Figure 12.
Biplot PCA multivariate analysis: TP, TN, and TN/TP ratio.

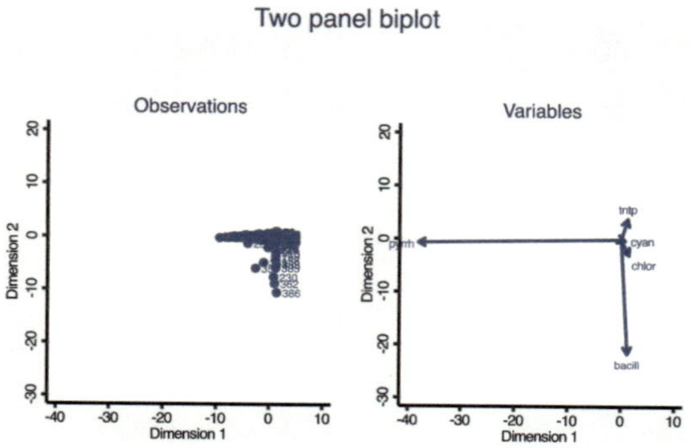

Table of Biplot Coordinates

Variable.	Dimension 1.	Dimension 2
TN/TP.	1.2531.	3.9121
Cyanophyta	0.7221	-0.2367
Bacillariophyta.	1.1948	-21.7066
Chlorophyta	1.4679	-2.9906
Pyrrhophyta	-38.1910	-0.6702

Figure 13.
Biplot PCA multivariate analysis.

nitrogen-fixing cyanophytes gained public attention. As in many other lakes in the Northern Hemisphere, predictive precaution was confirmed during the summer of 1994 in Lake Kinneret, when a bloom of *AO* was documented. A retrospective evaluation of the preliminary information about the TN/TP ratio in the Kinneret epilimnion (**Figures 5–8**) justified those preconceived warning symptoms.

Results of PCA analysis, shown in **Figure 12**, indicate that among TP, TN, and TN/TP variables, the TN is a principal component.

TN/TP Mass Ratio, Cyanophyta, Bacillariophyta, Chlorophyta, Pyrrhophyta.

Results of PCA analysis, shown in **Figure 13**, indicate that among Cyanophyta, Bacillariophyta, Chlorophyta, Pyrrhophyta, and TN/TP variables, the Pyrrhophyta (*Peridinium*) and Bacillariophyta are principal components, and the decline of Pyrrhophyta and Bacillariophyta is prominently distinct.

3.7 Long-term (1970–2018) temporal changes of nutrient concentrations in the Jordan waters

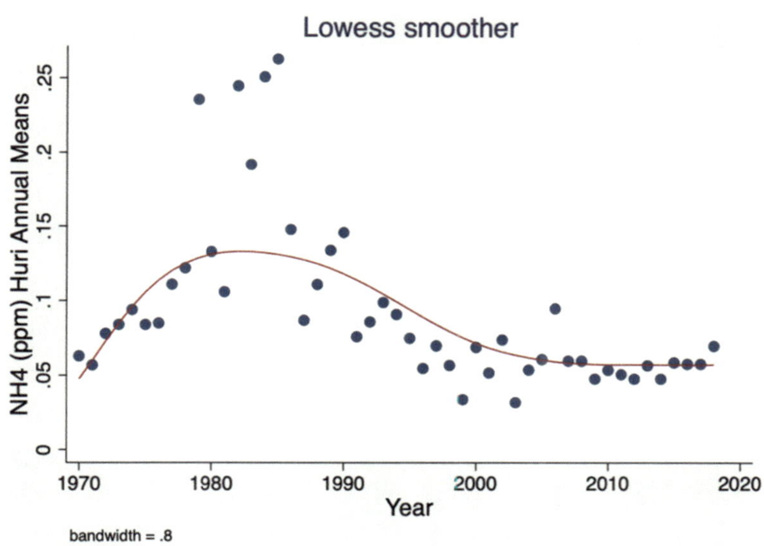

Figure 14.
Lowess smoother plot of temporal (1970–2028) changes in monthly means of NH_4 concentration (ppm) measured in the River Jordan (Huri Station).

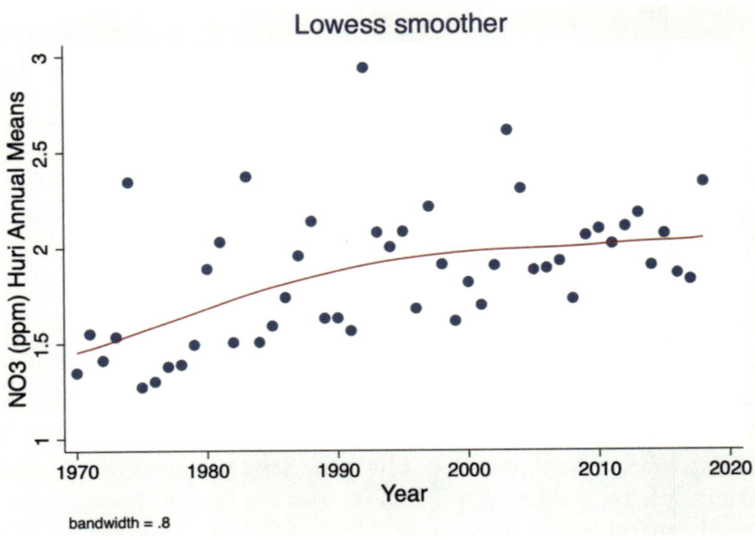

Figure 15.
Lowess smoother plot of temporal (1970–2028) changes in monthly means of NO_3 concentration (ppm) measured in River Jordan (Huri Station).

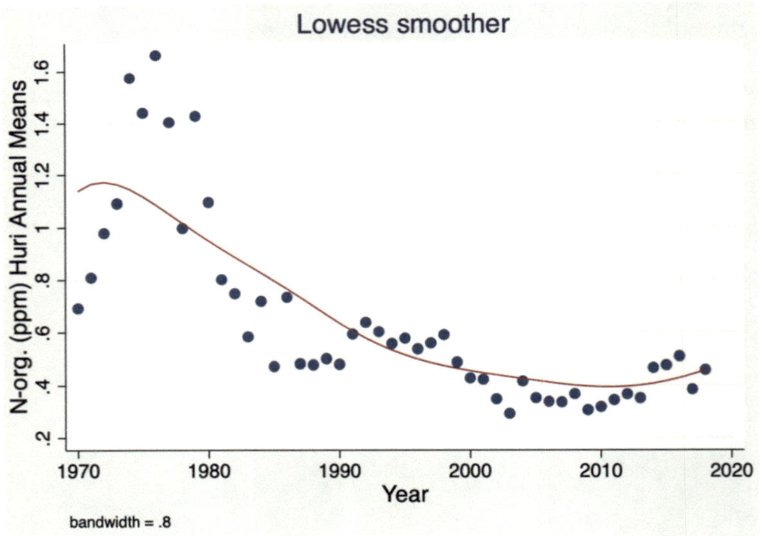

Figure 16.
Lowess smoother plot of temporal (1970–2028) changes in monthly means of N-org concentration (ppm) measured in the River Jordan (Huri Station).

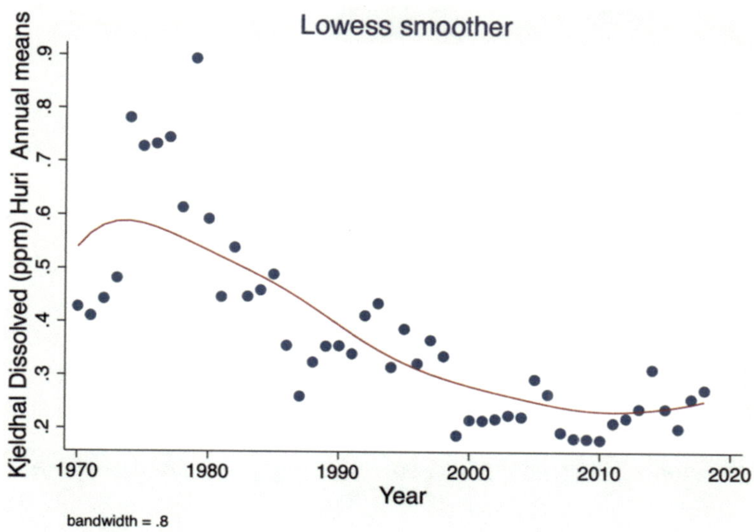

Figure 17.
Lowess smoother plot of temporal (1970–2028) changes in monthly means of Kjeldahl dissolved concentration (ppm) measured in River Jordan (Huri Station).

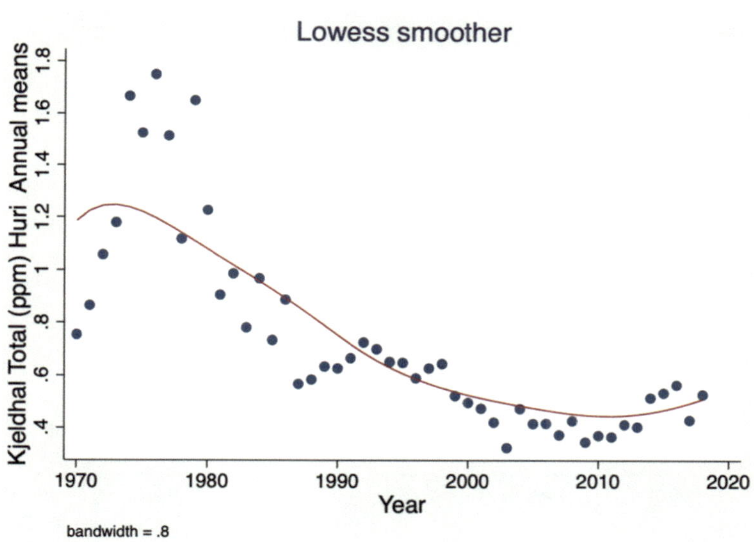

Figure 18.
Lowess smoother plot of temporal (1970–2028) changes in monthly means of Kjeldahl total concentration (ppm) measured in the River Jordan (Huri station).

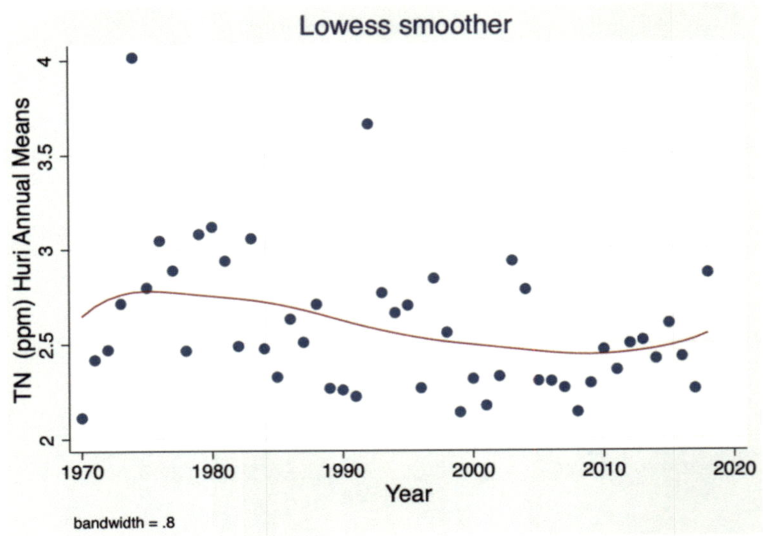

Figure 19.
Lowess smoother plot of temporal (1970–2028) changes in monthly means of TN concentration (ppm) measured in River Jordan (Huri station).

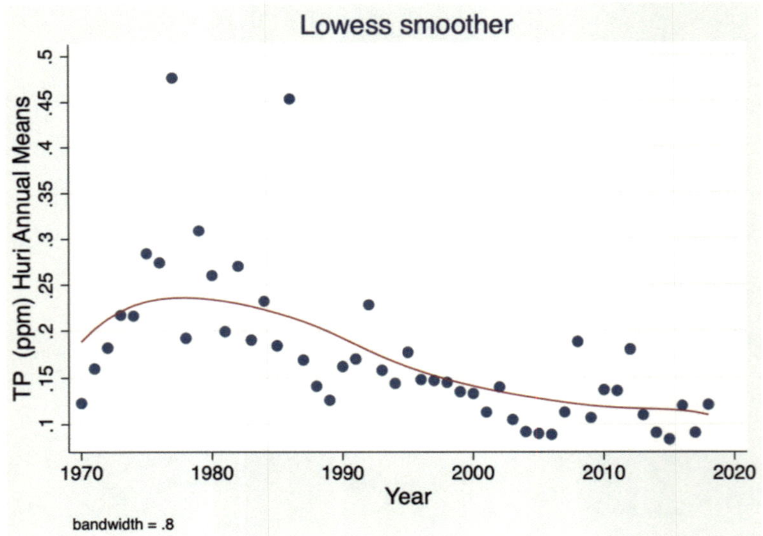

Figure 20.
Lowess smoother plot of temporal (1970–2028) changes in the means of TP concentration (ppm) measured in the River Jordan (Huri Station).

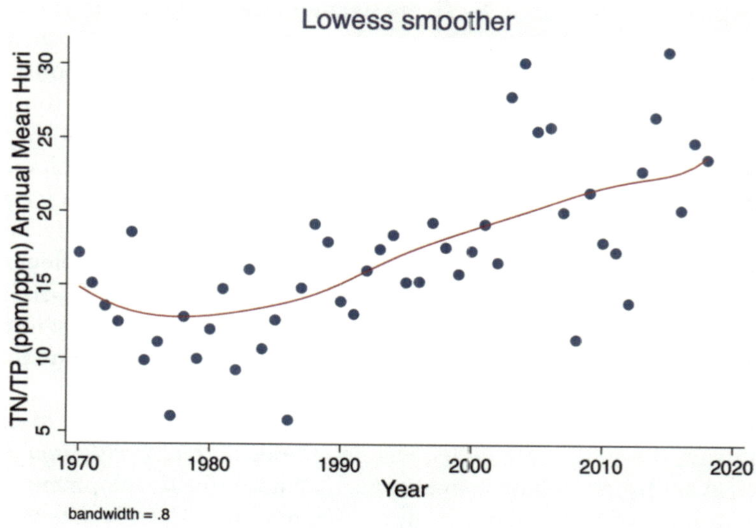

Figure 21.
Lowess smoother plot of temporal (1970–2028) changes in the means of TN/TP ratio concentration measured in River Jordan (Huri station).

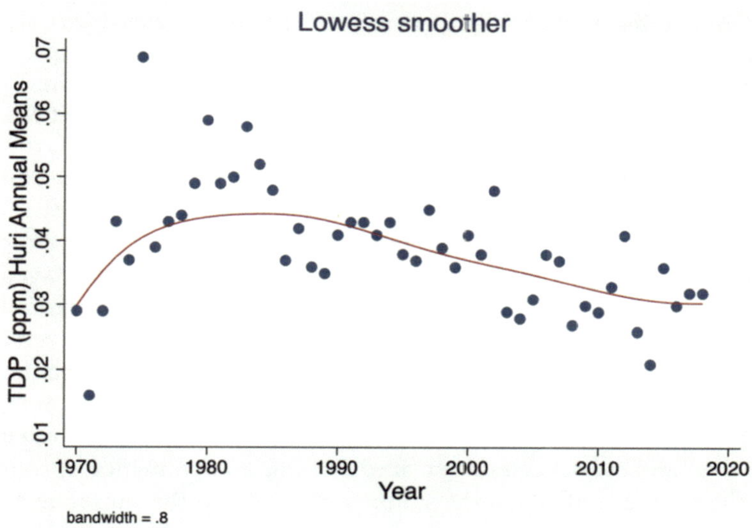

Figure 22.
Lowess smoother plot of temporal (1970–2028) changes in monthly means of TDP concentration (ppm) measured in River Jordan (Huri Station).

4. Discussion

Terrestrial and aquatic, particularly in Lake Kinneret, cyanobacteria are widely distributed organisms in Israel [27–32]. Despite a continuous long-term record of Cyanophyta, in recent decades, it has not been a source of substantial concern for drinking-water quality. This is because of their low biomass density and species diversity, which excluded toxic N_2-fixers, while random blooms of *Microcystis spp.* occurred in reservoirs and fishponds; no other cyanobacteria were recorded. The recorded short-term *Microcystis* blooms in Lake Kinneret were a response to Jordan River floods during early winter, which imported nitrate loads prior to the *Peridinium* blooming season [33]. Ecological sorting of cyanobacteria by properties includes, among others, single-cell or colonial, filamentous or round-shaped colonial types, toxic and non-toxic species, N_2-fixers and non-fixers, buoyancy capabilities, and vulnerability to consumer herbivory, and most crucially, response to nutrient regimes and composition. Each one of these single sorting types is relevant to the threatening level of water quality deterioration. Nevertheless, the N_2-fixing capabilities and toxicity of *AO and Cylindrospermopsis raciborskii have* initiated crucial concerns among water supply managers regarding the water quality of Lake Kinneret.

Suitable conditions for cyanobacteria, which gradually developed in Lake Kinneret during the 1980s and 1990s, enabled the decline of *Peridinium* and encouraged cyanobacteria enhancement. Furthermore, the development of nitrogen deficiency, accompanied by phosphorus sufficiency, significantly affected certain species, particularly HFCB such as AO. The most noticeable change in the Kinneret ecosystem was the modification of the nitrogen and phosphorus input loads and, consequently, their epilimnetic availabilities. Before the Hula drainage, a major part of the external nitrogen supply was in the form of ammonium. The algal affinity for this form of nitrogen is high and was favored by the bloom-forming *Peridinium*. After the Hula drainage, the external supply of ammonium was drastically reduced, and the dominant nitrogen form became nitrate. However, the prompt implementation of agricultural management was achieved through the establishment of 1,700 hectares of fishponds. The typically low winter temperatures required continuous headwater flows through the aquaculture ponds, resulting in external nutrient-mediated loads of high-quality headwater being introduced into Lake Kinneret. Furthermore, heavy nutrient loads mediated by domestic and dairy raw sewage effluents were supplied to the lake. Consequently, sewage and fishpond effluents replaced the pre-Hula drainage conditions that were advantageous for ammonium. Radical and far-reaching management measures were carried out, including sewage removal and fishpond restrictions, with an 80% conversion to field crops [21]. Despite nitrate migration from the oxidized peat soil, organic nitrogen input was reduced, and nutrient conditions in the Kinneret epilimnion were modified [28]: The N/P mass ratio declined, phosphorus replaced nitrogen as the limiting factor, and cyanobacteria replaced *Peridinium* dominance. It soon became clear that the outbreak of AO in the summer of 1994 and the subsequent HFCB blooms were not one-time phenomena. Future predictions of repeatable outbreaks of HFCB had far-reaching consequences for the design of Lake Kinneret's management strategies.

It is widely known that the removal or reduction of external nutrient inputs reduces or prevents lake eutrophication [34]. However, the exceptional case of the Kinneret ecosystem confirmed that external nitrogen removal without a simultaneous decline in phosphorus prevented eutrophication in the usual common sense, but water quality deteriorated through HFCB enhancement. The Kinneret ecosystem has far-reaching potential for maintenance against outbreaks of toxic and non-toxic, N_2-fixing and non-N_2-fixing cyanobacteria. Management design aimed at ecosystem improvement includes phosphorus reduction by enhancement of summer irrigation in the Hula Valley and renewal of the hydrological system structure aimed at bypass design for the diversion of peat soil rainwater directly toward the lake. This is because there is a nitrogen decline (denitrification and sedimentation) within the LAH ecosystem.

The outbreak of AO was recorded at 2,500 and 1,000 filaments/cm^3 during the summers of 1994 and 1995, respectively [12]. From 1994 onwards, toxic N_2-fixing cyanobacteria (HFCB) were sporadically recorded. Fluctuating densities of the toxic, non-N_2-fixing *Microcystis spp.* colonies, spread over most of the lake surface, were recorded every year. The existence of microcapsules within the AO cells enables buoyancy. The altitude of the National Water Carrier (NWC) intake is 215 meters below sea level. From there, water is pushed upward to 5 meters above sea level, that is, an altitude difference of 220 meters along a horizontal distance of about 30 meters – resulting in a 20-atmosphere pressure difference. Though the created pressure difference induces the collapsing of the intracellular capsules and cancellation of the buoyancy capabilities, enabling removal through sedimentation traps [35], toxicity remains despite buoyancy cancellation. The level of *Microcystins* toxicity is routinely sampled at 7–15 stations throughout the entire lake surface area, contributed by *Microcystis spp*. bloomed aggregation is monitored. Results indicate concentrations below the permitted level in the NWC intake lake zone. The herbivore zooplankton gut content was studied [1], and a low level of consumption of cyanobacteria, along with the predominance of Chlorophyta and Bacillariophyta, was confirmed.

Prior to the record of *AO* in 1994, more than 40 Cyanophyte species were documented in Lake Kinneret [27]. Of these, only one genus (*Microcystis spp.*) is HFCB but is a non-N_2-fixer. Since the late 1960s, the Lake Kinneret ecosystem has undergone significant external and internal environmental changes: Shifting from phosphorus to nitrogen as the limiting nutrient, which led to the replacement of *Peridinium spp*. by Cyanophyta-dominated blooms. The predictive capabilities of earlier ecosystem models [1, 2] were evaluated retrospectively.

Since the early 1930s, three critical management achievements were implemented within the lake and its drainage basin ecosystems: The construction of the south dam (1933), which enabled human control of Lake Water Level Fluctuation (WLF); the construction of the NWC (1964), which established Lake Kinneret as a national source of domestic water supply through a daily withdrawal of 10^6 m^3; and the Hula drainage (1957), which enhanced nitrate migration into the lake. The commonly accepted limnological paradigm of lake and reservoir ontogeny suggests that water-level fluctuations may increase sedimentary nutrient release, causing long-term eutrophication [36, 37]. Consequently, eutrophication is attributed to the enhancement of nutrient loads. After the Hula drainage was completed and nitrate migration was enhanced, national concern about Kinneret eutrophication increased. A distinct decline in nitrogen (**Figure 6**) and a slight increase in phosphorus loads (**Figure 7**) in the epilimnion were documented. Taking into account nitrogen availability as the leading factor, water quality in Lake Kinneret improved, while, due to the long-term minor elevation of phosphorus and the N/P mass ratio decrease (**Figure 8**), water quality deteriorated. The desired algal community composition, dominated by *Peridinium spp*. biomass in Lake Kinneret indicates better water quality [36], whereas the enhancement of Cyanophyta, including HFCB (*Aphanizomenon spp. and Cylindrospermopsis spp.*), indicates water quality deterioration. The ecosystem-modulated evaluation of the Kinneret ecosystem [1, 2, 36, 37] did not indicate model predictions of optional developments suitable for the enhancement of cyanobacterial biomass. None of these documents formulated a clear, distinct precaution or an optional predominant alarm signal emphasizing the replacement of *Peridinium* by Cyanophyta in Lake Kinneret. These scientifically modulated or "state-of-the-art" brilliant evaluations are more eco-limnological "Post Mortem Autopsy" than valuable management predictions. Despite the predicted management design of unwanted fish removal, it was concluded as a crossword clue [1, 37].

The long-term background of nutrient dynamics, which created those ecosystem modifications, was widely, partially, and comprehensively documented. Several models were suggested [1, 2, 37], whereas none of them presented a prediction of the management design of the watershed nutrient regime resulting from respective modulated nutrient dynamics. Furthermore, forecasted optional modifications of nutrient availabilities that support the enhancement of cyanobacterial blooms are not considered as thoroughly as they are for zooplankton.

Model A [1] considered two types of phytoplankton communities that characterized different eco-limnological conditions: A) Low density of Pyrrhophyta accompanied by enhanced biomass of non-Pyrrhophyte tiny algae components, mostly Chlorophyta, Bacillariophyta, and nontoxic non-colonial nanoplanktonic Cyanophyta. B) The commonly recorded late-winter-spring-early-summer heavy bloom of *Peridinium gatunenze* and low density of nanophytoplankton. In both cases, the summer phytoplankton assemblages are dominated by nanophytoplankton, of which Cyanophyta biomass is low. The *Peridinium* bloom is characterized by high availability of nitrogen (mostly as nitrate), starting shortly before bloom initiation, resulting from heavy external inputs. Independently, the biomass enhancement of *Peridinium* after bottom excystation and the pelagic high rate of vegetative cell multiplication is also abruptly supported by external and internal epilimnetic phosphorus enrichment elevation. Four channels of phosphorus input into the Kinneret epilimnion are indicated: External aquatic headwaters, atmospheric dust deposition, internal flux from bottom sediments through mediated *Peridinium* cells, and geochemical-bacteriological processes. As soon as the nitrogen supply is overwhelmed, the *Peridinium* creates a flourishing bloom, which, by itself, is a consequent enhancement of phosphorus input into the epilimnion. Nevertheless, from the late 1980s to early 1990s, external aquatic input of nitrogen diminished, but bottom phosphorus supply did not decline, supporting epilimnetic nutrient availability conditions. Though the epilimnetic nutrient balance was modified into "Sufficient Phosphorus and Insufficient Nitrogen," which is preferably favored by cyanobacteria and unfavored by *Peridinium*. The interaction between nutrient composition, such as the mass ratio between the availabilities of nitrogen and phosphorus, was not considered as a management factor in previous studies [1, 2, 36, 37]. The appropriate evaluation of the ecological processes that supported a significant change in the algal community includes long-term analysis. Statistical analysis of long-term precipitation (mm/year; number of rainy days/year) records (Dafna Station, 1969–2001) indicates no significant fluctuations. Consequently, other environmental factors within the ecosystem, such as anthropogenic management changes in the Hula Valley, which affected the nutrient input regime (**Figure 7**), were investigated. Consequently, temporal TN and TP dynamic fluctuations are dissimilar due to the different management impacts on TN and TP migration from the Hula Valley, as well as unequal rates of internal processes resulting in changes in the variance of the mass ratio between them (**Figure 8**).

4.1 The uniqueness of AO blooming in 1994

The epilimnetic section (0–15 m) of the Kinneret ecosystem during summer time is similar to a bio-chemostat ecosystem: High temperature leading to metabolic range demands enhancement of all living organisms from planktonic algae and bacteria through protozoans to metazoan invertebrates (zooplankters) and vertebrates (fish). On the other hand, headwater water and nutrient inputs and consequently their epilimnetic availabilities, strictly diminished. During summer months (August–October), the epilimnetic concentration of inorganic nitrogen ranged between 100–50 μg/l. In lakes where the concentration of inorganic nitrogen is lower than 50 μg/l, the N_2-fixing cyanobacterium is very common [15, 16, 38]. Moreover, as a result of summer stable stratification, internal upward nutrient transfer from the hypolimnion to the epilimnion is limited despite of a high velocity daily west wind regime and significant motions of water mass. The ecological situation of limited food resources, accompanied by the enhancement of energy requirement, creates a sensitive seasonal ecosystem of the Kinneret epilimnion. Which require at least conditions supply of the following: Insufficient nitrogen, sufficient phosphorus (INSP), resulting in a lower N/P ratio, and appropriate inoculum. Nitrogen limitation interferes with non-N_2 fixing algal metabolic activity; sufficient phosphorus is required by the entire algal assemblage, including nitrogen fixers. The minimum sufficiency inoculum demand is due to the short period for the appropriate combined conditions to exist. The point-specific Kinneret summer conditions competitive advantage of the N_2-fixer cyanobacterium is their ability to utilize atmospheric nitrogen by the nitrogenase enzyme. Despite the short-term suitable niche for AO bloom formation during summer 1994, the long-term perspective confirms the N/P mass ratio and INSP availability for cyanobacteria and HFCB, including enhancement and consequently replacing *Peridinium* dominance unless epilimnetic nitrogen load is significantly enriched. During the 15 years prior to the AO bloom outbreak, epilimnetic nitrogen deficiency was progressively developed (**Figure 6**). Wind velocity record indicates exceptionally low regimes during the summers of 1994 and 1995. It is a factor of mixing water mass reduction, which diminished the upward transfer of inorganic nitrogen as well. The long-term decline of nitrogen, whilst phosphorus was enhanced, resulting N/P mass ratio decrease and the AO outbreak were early warning signals of the cyanobacterial threat to water quality. The *Peridinium* decline and blue-green replacement, which came later, confirmed this trend, but not necessarily eutrophication and dissolved oxygen depletion. Moreover, the proliferation of nitrogen fixers contributes enrichment of nitrogen load (supporting the renewal of *Peridinium* blooming) as already measured during the AO summer bloom.

Results given in **Table 4** and **Figures 15–22** emphasize the role of sewage removal, fishpond restriction, and replacement by field crops in the achievements of the post-drainage Hula Valley on nitrogen and phosphorus migration. These management operations highlight the potential role that nitrogen and phosphorus input loads play in cyanobacteria dynamics. Nitrogen forms supplied from the Hula wetlands to the Kinneret ecosystem before 1957 were mostly ammonia and organic nitrogen [22]. The Hula drainage and the consequent peat soil oxidation process caused a shift in the migrated nitrogen composition from ammonia to a nitrate-dominated content. A major part of the drained Hula soils (peat) was organic-nitrate-rich and weakly bound to the soil particles. Practically, the upper soil layers of the Hula post-drainage were created as a huge bulk of easily flushed inorganic nitrogen – nitrate [25]. Unlike the anthropogenic impact on the decline of organic nitrogen and phosphorus, nitrate migration, and consequently TN (**Figures 6, 19**) is mostly affected by soil moisture contributed by rainfall and summer irrigation capacities. Therefore, a gradual temporal enhancement of nitrate concentration in the Jordan waters was recorded (**Figure 15**). The decline in phosphorus concentration in the Jordan waters, supported by management (**Figures 7, 20**) combined with nitrate enhancement, affected the TN/TP ratio elevation (**Figures 8, 21**). Despite the decline in external phosphorus supply, an increase in epilimnetic concentration was recorded during the mid-1990s, supported by internal sources [11, 12]. Because nitrate is the dominant nitrogenic-component within the Jordan waters, the TN concentration did not decline (**Figure 19**). The affinity of ammonia within the metabolic requirements of *Peridinium* is high, resulting, probably, in its heavy bloom existence before drainage and shortly (mid-1990s) later when cyanobacteria enhanced. Despite peat soil oxidation and nitrate migration enhancement (**Figure 15**), ammonia and organic nitrogen concentrations were severely enhanced as a result of sewage and fishpond effluents (1970–1985: **Table 4, Figures 14, 16–18**). Sewage and fishpond effluent inflows into the Jordan discharge also enhanced the migration of phosphorus (**Figures 20, 22; Table 4**). The enhancement of organic nitrogen, ammonia, and phosphorus loads (1970–1985) through sewage and fishpond effluents (1970–1985) was later (1986–2018) declined, excluding nitrate. Consequently, total nitrogen concentration in the Jordan waters was fairly stable (**Figure 19**) whereas a significant decline in organic nitrogen and ammonia occurred as a result of sewage and fishpond effluent management removal. The *Peridinium* decline and cyanobacteria enhancement were initiated during the mid-1990s despite nitrate availability, taking into account the higher affinity of organic nitrogen and ammonia to *Peridinium* growth than nitrate. With respect to external Kinneret inputs of nitrogen and phosphorus, shortly after drainage (1970–1986), conditions achieved *Peridinium* dominance, whilst later on cyanobacteria were favored: Nitrogenic forms (organic and ammonia) deficiency and phosphorus (internal and external) sufficiency. The enhancement of the TN/TP ratio (**Table 4, Figure 21**) is due to the enhancement of nitrate and consequently TN, whilst the cyanobacteria-preferred nitrogen forms (NH4, organic – N) declined. The decline of the TN/TP mass ratio, resulting from nitrogen deficiency and phosphorus sufficiency in the epilimnion of Lake Kinneret, is widely documented [11, 15, 16, 21] and enhanced N_2-fixing cyanobacterium. The intra cellular enabled nitrogen-fixer AO to flourish during the summer of 1994 in Lake Kinneret.

5. Conclusive summary

The seasonal respective response of nutrient input with Jordan discharge is shown in **Figures 1–4**. The dependence of epilimnetic loads of nitrogen and phosphorus on the external inputs is emphasized in **Figures 5–8**. Further consideration implies theresulting cyanobacteria enhancement and *Peridinium* decline, reinforced by the PCA analysis (**Figures 9–13**). Justification of the short-term (1970–2001) changes' similarity to the long-term (1970–2018) period is confirmed in **Figures 14** and **16–22**. The significant impact of anthropogenic involvement within the Hula Valley's ecological structure is limited to sewage removal and the conversion of fishponds to field crops. On the other hand, the Hula drainage caused peat soil oxidation and modification of migrated nitrogen from ammonia before drainage to nitrate after drainage. The very high peat nitrogen content and mobility through wettability imply its continuous temporal increase (**Figure 15**). The focused concept of the paper is aimed at the management implementation within the Hula Valley after drainage.

During the late 1960s to 1990s, regular heavy late winter-spring-early summer blooms of *Peridinium* were recorded. Then, *Peridinium* densities declined, and cyanobacteria increased. The evaluated study of the long-term (1969–2001) spatio-temporal record of hydrological dynamics, combined with nutrient migrations from the Kinneret drainage basin and phytoplankton and nutrient distributions in Lake Kinneret, confirmed interactions between them. The data shown in **Figures 1–5** confirm that, in correlation with seasonal rainfall distribution, the epilimnetic loads of TP and TN are high in winter and decline in summer. The temporal data (1969–2001) (**Figures 6–8**) confirm a decline in TN and a slight increase in TP, and therefore, the TN/TP mass ratio in the epilimnion of Lake Kinneret declined. Low ratios of N/P are well-known as suitable conditions for Cyanophyta enhancement [15, 16]. The merit of nitrogen availability to *Peridinium* growth and multiplication is shown in **Figure 9**: the more nitrogen is available, the heavier the *Peridinium* biomass density. However, *Peridinium* biomass density also increases with the availability of TP in the epilimnion (**Figure 10**). The concluded outcome about nitrogen deficiency and phosphorus sufficiency as N_2-fixer cyanobacterium enhancement (**Figures 9, 10**) is therefore unlikely, whilst it is likely if TP and TN sources are indicated. The sources of the epilimnetic nitrogen are mostly external, through infrastructure, rocky basement erosion by headwater flows, and, to a lesser extent, through internal ammonium oxidation during stratification and destratification and N_2 fixation. Temporal stability and a slight increase in TP external supply through erosion and dust deposition were documented. Moreover, a significant part of epilimnetic phosphorus import is carried out by the vegetative, multiplicated bloom-forming *Peridinium*. A significant part of this transported phosphorus is due to previously incorporated phosphorus from the sediments by the settled *Peridinium* cysts into the sediments after cystation. During late winter and excystation occurrence, vegetative cells acquire motile ability by flagellum and migrate into the epilimnion, creating bloom onset by the high rate of cell division. In early summer (June–July), cell degradation is enhanced. Throughout the entire bloom existence, and particularly during bloom collapse, the release of inorganic dissolved phosphorus (DIN) into the epilimnion was documented [1, 11, 12, 27]. Conclusively, epilimnetic nitrogen decline resulted from external supply reduction and internal phosphorus flux enhancement by *Peridinium* cell transport, creating suitable conditions for a particular AO outbreak [10–12, 35]. Long-term depletion of nitrogen external supply, accompanied by stability of phosphorus inputs, predicts enhancement of cyanobacteria biomass at the expense of *Peridinium*. The outcome for future management is the reduction of external phosphorus input. Enhancement of external nitrogen input, as nitrate, might be implemented through an irrigation regime.

Acknowledgments

Sincere appreciation is given to the Kinneret Limnological Laboratory, Israel Oceanographic and Limnological Research Co. Ltd., Haifa, and the Migal-Scientific Research Institute in the Galilee, Hula Project Data Base, and Mekorot Water Co. Ltd. for the available data.

Funding

This research received no external funding.

Author contribution

The author directed data evaluation, presentation design, computerization, and the preparation of the original and final draft versions.

Data availability statement

The data presented in this study is available upon request from the corresponding author.

Author details

Moshe Gophen
Migal-Galilee Scientific Research Institute, Tel-Hai University Under Construction, Kiryat Shmona, Israel

*Address all correspondence to: gophen@migal.org.il

References

[1] Serruya C, Gophen M, Pollingher U. Lake Kinneret: Carbon Flow patterns and ecosystem management. Archive Hydrobiology. 1980;**88**(3):265–302

[2] Bruce LC, Hamilton D, Imberger J, Gideon G, Gophen M, Zohary T, Hambright KD.A numerical simulation of the role of zooplankton in C, N and P cycling in Lake Kinneret, Israel. Ecological Modelling. 2006;**193** (3-4):412–436

[3] Gophen M. Chapter 11. Long-term (1970-1990) whole lake biomanipulation experience: Case study Lake Kinneret (Israel). In: De Bernardi R, and Giussani G, editors. Guidelines of Lake Management, Vol. 7: Biomanipulation in Lakes and Reservoirs Management. International Lake Environment Committee, United Nations Environment Programme. Rome: United Nations Publications; 1995. pp. 171–184

[4] Gophen M. The management of Lake Kinneret (Israel): Water supply, water quality and food web structure perspectives. Proceedings (Volume 2) of The Water and Environmental Research ICWRER 2002 (Schmitz GH, ed.), Drezden, Germany: ICWRER Publications; 2002. Vol. **2** pp. 275–285

[5] Gophen M. Hydrology and management of Lake Kinneret aimed at water quality protection. In Zereiny F, and Jaeschke W, editors. Water in the Middle East and in North Africa: Resources, Protection, and Management. Drezden, Germany: Springer-Verlag; 2003. pp. 207–226

[6] Gophen M.Ecohydrological management of Lake Kinneret: A case study. Ecohydrology & Hydrobiology. 2004;**4**(4):397–408

[7] Serruya CUP, Cavari BZ, Gophen M, Landau R, Serruya S. Lake Kinneret: Management Options. Archiv Fur Hydrobiologie Beihefte Ergebn Limnology. 1979;**13**:306–316

[8] Gophen M, and Serruya S. Effects of fish and nutrients on long-term changes of plankton communities in Lake Kinneret (Israel). In Schroder JH, Bauer J, Schartl M, editors. Trends in Ichthyology, an International Perspectives. Munich, Germany: Blackwell Scientific Publications; 1993. pp. 371–380

[9] Gophen M. Long-term Changes of Plankton-Communities in Lake Kinneret, Israel. Asian Fisheries Science. 1992;**5**:291–302

[10] Gophen M. Prediction of an N2-fixing cyanobacteria bloom in Lake Kinneret (Israel) based on N and P balance. Verhandlungen - Internationale Vereinigung Fur Theoretische Und Angewandte Limnologie. 1999;**27**:1239–1242

[11] Gophen M, Smith VH, Nishri A, Threlkeld ST. Nitrogen deficiency, phosphorus sufficiency, and the invasion of Lake Kinneret, Israel, by N_2-fixing cyanobacterium *Aphanizomenon ovalisporum*. Aquatic Sciences. 1999;**1**:1–14

[12] Hadas O, Pinkas R, Delphine E, Vardi A, Kaplan A, Sukenik A. Limnological and ecophysiological aspects of *Aphanizomenon ovalisporum* bloom in Lake Kinneret, Israel. Journal of Plankton Research. 1999;**21**(8): 1439–1453

[13] Gophen M. Experimental study of the *Aphanizomenon ovalisporum* response to Nitrogen fertilization in the Sub-Tropical Eu–Meso-trophic Lake Kinneret (Israel). Open Journal of Ecology. 2015b;**5**:259–265. DOI: 10.4236 /oje.2015.56020.

[14] Gophen M. The Impact of Nitrogen and Phosphorus Dynamics on the Kinneret Phytoplankton: I: Cyanophytes-Peridinium Alternate. Open Journal of Modern Hydrology. 2017a;**7**:257–273. DOI: 10.4236/ ojmh.2017.74015.

[15] Smith VH. Low Nitrogen to Phosphorus ratios favor dominance of Cyanobacteria in lake Phytoplankton. Science. 1983;**221**:669–671

[16] Smith VH. Light and nutrient effects on the relive biomass of blue-green algae in lake Phytoplankton. Canadian Journal of Fisheries and Aquatic Sciences. 1986;**43**: 148–153

[17] Gophen M. The impact of available nitrogen defficiency on long-term changes in the Lake Kinneret Ecosystem. Open Journal of Ecology. 2015a;**5**:147–157. DOI: 10.4236/ oje.2015.54013.

[18] Gophen M. The impact of nitrogen and phosphorus dynamics on the Kinneret Phytoplankton: I: Cyanophytes-Peridinium alternate. Open Journal of Modern Hydrology. 2017a;**7**:257–273. DOI: 10.4236/ ojmh.2017.74015.

[19] Gophen M. The impact of Nitrogen and Phosphorus Dynamics on the Kinneret Phytoplankton: II: Chlorophyta, Cyanophyta Diatoms and Peridinium. Open Journal of Modern Hydrology. 2017b;**7**:298–313

[20] Gophen M. Limnological changes in the Lake Kinneret (Israel) Ecosystem: The beginning and its significance. Open Journal of Modern Hydrology. 2016;**6**:140–150. DOI: 10.4236/ ojmh.2016.63012.

[21] Gophen M. Nitrogen and phosphorus removal from Lake Kinneret Inputs. Open Journal of Ecology. 2024; **14**: 165–182. https://www.scirp.org/journal/oje.

[22] Dimentman C, Bromely HJ, Por FD. Lake Hula, reconstruction of the Fauna and Hydrobiology of a Lost Lake. Publications of the Israel Academy of Sciences and Humanities. 1992 Brief Summary in Hebrew 24p.;**170**:

[23] LKDB (2022) 1969-2001, phytplankton density and nutrient concentration distribution (Courtesy: M. Schlichter, A. Nishri) and Annual Reports. Kinneret Limnological Laboratory, IOLR, Ltd.

[24] Gophen M. Experimental study of the *Aphanizomenon ovalisporum* response to Nitrogen fertilization in the Sub-Tropical Eu–Meso-Trophic Lake Kinneret (Israel). Open Journal of Ecology. 2015b;**5**:259–265. DOI: 10.4236 /oje.2015.56020.

[25] Gophen M. Seasonal and Spatio-Temporal Distribution of Nutrients in the Hula Valley after Drainage: C: Carbonate and Sulfate. Open Journal of Modern Hydrology. 2025;**15**:204–217. DOI: 10.4236/ojmh.2025.152013.

[26] Serruya C. Chapter:C: Sedment Chemistry. In Serruya C, editor. LAKE KINNERET Monographiae Biologicae, Vol. 32. The Hague -Boston-London1978: Dr. W. Junk bv Publishers; 1978. pp. 205–215

[27] Pollingher U. Chapter: The Phytoplankton of Lake Kinneret. In Serruya C, editor. LAKE KINNERET Monographiae Biologicae, Vol. 32. The Hague -Boston-London1978: Dr. W. Junk bv Publishers; 1978. pp. 229–243

[28] Geifman Y, Shaw M, Dexster J, Sarusi F. Research of nutrien lads in jordan river and headwaters: 1968-1985. Jordan District, Kinneret Watershed Unit: Data Volume Attachments; Mekorot, Water Co. Ltd; March 1987. p. 155. (in Hebrew).

[29] Rayss T. Materiaux pour la flore Algologique de la Palestine: Les Cyanophycees. Plastic Journal of Botany Jerusalem. 1944;**3**:94–113

[30] Rayss T. Materiaux pour la flore Algologique de la Palestine: II Les Algues des eaux continentals. Plastic Journal of Botany Jerusalem. 1951;**5**:51–95

[31] Dor I. Algues de sourses thermals de Tiberiade. Sea Fish Research State Bulletin. 1967;**48**:23–129

[32] Vinogradova ON, Kovalenko OV, Wasser SP, Nevo E. Cyanophyta: Checklist of Continental Species from Israel, International Center for Cryptogamic Plants and Fungi, Institute of Evolution University of Haifa. National Academy of Sciences of Ukraine. Haifa-Kyiv-1996: M.G. Kholodny Institute of Botany; 1996. p. 105

[33] Serruya C, Pollingher U. An attempt at forcasting the Peridnium bloom in Lake Kinneret (Lake Tiberias). Mitteilungen Internationale Vereinigung Limnology. 1971;**19**: 277–291

[34] Edmondson WT. The Uses of Ecology: Lake Washington and Beyond. Seattle and London: University of Washington Press; 1991. p. 331

[35] Porat R 1995, Aphanizomenon in Lake Kinneret and National Water Carrier; Interim Progress Report; Nazareth, Jordan Region: Mekorot, National Water Supply Company Ltd. Central Water Quality Laboratory; p. 46

[36] Gal G, Hipsey MR, Parparov A, Wagner U, Makler V, Zohary T. Implementation of ecological modeling as an effective management and investigation tool: Lake Kinneret as a case study. Ecological Modelling. 2009;**220**(13–14):1697–1718

[37] Hambright KD, Eckert W, Leavitt PR, Schelske CL. Effects of Historical Lake Level and Land Use on Sediment and Phosphorus Accumulation Rates in Lake Kinneret. Environmental Science and Technology. 2004;**38**:6460–6467

[38] Horne AJ, Commins ML. Macronutrient controls on nitrogen fixationin planktonic Cyanobacterial populations. New Zealand Journal of Marine & Freshwater Research. 1987;**21**:413–423